Balkan Mathematical Olympiads

the first 30 years (1984-2013)

Balkan Mathematical Olympiads

the first 30 years (1984-2013)

Mircea Becheanu

Bogdan Enescu

Library of Congress Control Number: 2014900360

ISBN-13: 978-0-9885622-5-7 ISBN-10: 0-9885622-5-1

9 8 7 6 5 4 3 2 1

www.awesomemath.org

Cover design by Iury Ulzutuev

Preface

Hundreds of mathematical competitions are organized these days all around the world. Almost every country holds its national Olympiad, and in some of them are town or regional competitions[1]. Also, there are regional competitions, such as the Balkan Mathematical Olympiad (and its younger sister the Junior Balkan Mathematical Olympiad), Asian Pacific Mathematical Olympiad, Baltic Way, Nordic Mathematical Contest, South East Asian Mathematics Competition, etc.

The International Mathematical Olympiad, queen of all math competitions, started in 1959, in Romania, with 7 attending countries. In 2009, in Germany, students from 104 countries were fighting for a medal.

The number of those interested in math competitions is constantly increasing and the popularity of websites like www.artofproblemsolving.com (a math forum with over 130000 members) is a proof for this.

It all started in 1894, in Hungary, when the Eötvös Competition, a math contest for secondary school students, was held for the first time. The competitors were given four hours to solve three problems individually (almost the same happens today at the International Mathematical Olympiad).

In the neighboring Romania, the first issue of the monthly Gazeta Matematica was published in 1895. The journal organized a competition for school students, which improved in format over the years and eventually became the Romanian National Mathematical Olympiad.

A math competition was first held in Sankt Petersburg, Russia in 1934 and in Poland in 1947. The Mathematical Association of America organized a competition for Metropolitan New York in 1950 and extended this to the entire country in 1957.

In the last decade of July 1959 students from Bulgaria, Czechoslovakia, German Democratic Republic, Hungary, Poland, Romania, and Soviet Union gathered together in Romania to compete in the first International Mathematical Olympiad. Ever since, the IMO has developed a rich legacy and has

[1]In Romania, for instance, there are around 50 regional competitions every year, each involving hundreds of students.

next year, another math competition, the Balkan Countries Mathematical Olympiad.

Although the rules of IMO are very encouraging for the students, given that approximately half of them can win prizes, the competition is difficult and many students come to be disappointed for different reasons. Therefore, a preliminary competition was considered very helpful. The aims of the Balkan Mathematical Olympiad (BMO) include:

a. challenge, encouragement, and development of mathematically gifted school students in all participating countries;

b. fostering friendly relationships among students and teachers of the member countries;

c. creation of opportunities for the exchange of information on school syllabi and practice within the member countries;

d. gaining experience and preparation for the IMO.

The first Balkan Mathematical Olympiad was organized in Athens, Greece, in 1984. The participating countries were Bulgaria, Greece, and Romania. The rules of the competition were approximately the same as the IMO's. The competition extended since, and 11 countries are nowadays official members of the contest, the list being not closed. In the last years, other teams such as Hungary, United Kingdom, Kazakhstan, France, Italy, Saudi Arabia, etc. took part as invited countries.

It is important to mention that BMO problems are usually original, but less difficult than in the IMO's. Therefore, many young and/or less experienced students are encouraged to solve them. Even more, doing this successfully these students are motivated to involve themselves more in mathematics.

The authors of the book, attending several times the BMO as leader and/or deputy leader, present to the readers a complete description of the evolution of BMO's since their creation up to today. All problems are presented with complete solutions. Many problems have several alternative solutions and we also present some extensions. An additional preparatory addendum, containing concepts and classical useful results has been added to the end of the book.

The Authors

Contents

Part I

Problems and Solutions

The 1st BMO

The first Balkan Mathematical Olympiad for high-school students was held between May 6th and May 10th , 1984, in Athens, Greece. The competition was organized by the Greek Mathematical Society. The participating countries were Bulgaria, Greece, and Romania, the founder countries of this competition.

Problems

1.1. Let $n \geq 2$ be a positive integer and $a_1, a_2, \ldots, a_n$ be positive real numbers such that $a_1 + a_2 + \ldots + a_n = 1$. Show that the following inequality holds:

$$\frac{a_1}{1 + a_2 + a_3 + \ldots + a_n} + \frac{a_2}{1 + a_1 + a_3 + \ldots + a_n} + \ldots + \frac{a_n}{1 + a_1 + \ldots + a_{n-1}} \geq \frac{n}{2n-1}.$$

(Greece)

1.2. Let $ABCD$ be a cyclic quadrilateral and let H_A, H_B, H_C, H_D be the orthocenters of the triangles BCD, CDA, DAB, and ABC, respectively. Show that the quadrilaterals $ABCD$ and $H_A H_B H_C H_D$ are congruent.

(Romania)

1.3. Show that for any positive integer m, there exists a positive integer n, so that in the decimal representations of the numbers 5^m and 5^n, the representation of 5^n ends in the representation of 5^m.

(Bulgaria)

1.4. Let a, b, c be positive real numbers. Find all real solutions (x, y, z) of the system:

$$\begin{aligned} ax + by &= (x - y)^2 \\ by + cz &= (y - z)^2 \\ cz + ax &= (z - x)^2 . \end{aligned}$$

(Romania)

Solutions

1.1. Since $\sum\limits_{k=1}^{n} a_k = 1$, the given inequality can be written as follows:

$$\sum_{k=1}^{n} \frac{a_k}{2-a_k} \geq \frac{n}{2n-1}. \tag{1}$$

Note that

$$\frac{a_k}{2-a_k} = \frac{2}{2-a_k} - 1,$$

therefore, inequality (1) is equivalent to

$$2\sum_{k=1}^{n} \frac{1}{2-a_k} - n \geq \frac{n}{2n-1},$$

or

$$\sum_{k=1}^{n} \frac{1}{2-a_k} \geq \frac{n^2}{2n-1}. \tag{2}$$

The latter follows easily from the HM-AM inequality: observe that $2-a_k > 0$, so that we have

$$\frac{n}{\sum\limits_{k=1}^{n} \frac{1}{2-a_k}} \leq \frac{1}{n}\sum_{k=1}^{n} (2-a_k) = \frac{1}{n}\left(2n - \sum_{k=1}^{n} a_k\right) = \frac{2n-1}{n}.$$

Second solution. We prove (1) by using Jensen's inequality.
Consider the convex function $f : (0,1) \to \mathbb{R}$, $f(x) = \dfrac{x}{2-x}$. Then we have

$$\frac{1}{n}\sum_{k=1}^{n} \frac{a_k}{2-a_k} = \frac{1}{n}\sum_{k=1}^{n} f(a_k) \geq f\left(\frac{1}{n}\sum_{k=1}^{n} a_k\right) = f\left(\frac{1}{n}\right) = \frac{1}{2n-1}.$$

Third solution. We prove (1) using Jensen's inequality and Cauchy-Schwartz inequality. Consider the convex function $f : (0,1) \to \mathbb{R}$, $f(x) = \dfrac{1}{2-x}$, and apply Jensen's inequality to the numbers $a_1, \ldots, a_n$, and weights $a_1, \ldots, a_n$, satisfying the condition $\sum\limits_{k=1}^{n} a_k = 1$. We obtain

$$\sum_{k=1}^{n} a_k f(a_k) \geq f\left(\sum_{k=1}^{n} a_k^2\right).$$

Explicitly, we have

$$\sum_{k=1}^{n} \frac{a_k}{2-a_k} \geq \frac{1}{2-\sum\limits_{k=1}^{n} a_k^2},$$

therefore, it is sufficient to prove that

$$\frac{1}{2-\sum\limits_{k=1}^{n} a_k^2} \geq \frac{n}{2n-1},$$

which is equivalent to

$$\sum_{k=1}^{n} a_k^2 \geq \frac{1}{n}.$$

This inequality can be obtained by using Cauchy-Schwartz inequality, as follows:

$$1 = \left(\sum_{k=1}^{n} a_k\right)^2 \leq (1+\ldots+1)\left(a_1^2+\ldots+a_n^2\right) = n\sum_{k=1}^{n} a_k^2.$$

1.2. Let O be the circumcenter of the quadrilateral $ABCD$, M be the midpoint of the segment AB and G_A, G_B, G_C, G_D be the centroids of triangles BCD, CDA, DAB, and ABC, respectively.

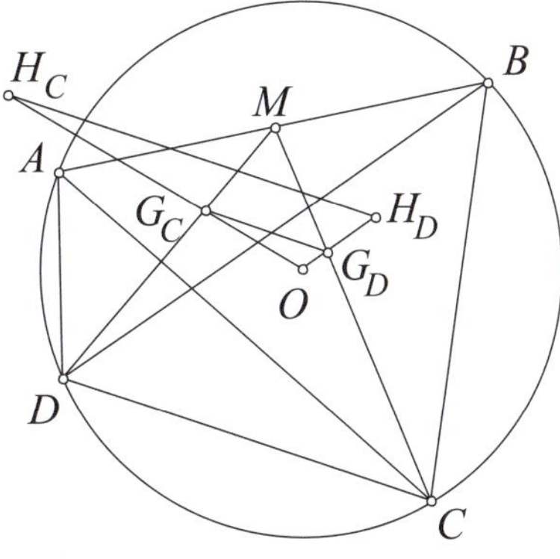

Figure 1.1

It is known that G_C lies on the segment DM and that $MG_C = \frac{1}{3}MD$. Similarly, G_D lies on the segment CM and $MG_D = \frac{1}{3}MC$. Therefore, in the triangle CMD, the segment G_CG_D is parallel to CD and $G_CG_D = \frac{1}{3}CD$ (see Figure 1.1).

On the other hand, it is known that in the triangle ABD, the orthocenter H_C, the centroid G_C, and the circumcenter O lie on the same line (the Euler line-see Appendix) in such a way that $OG_C = \frac{1}{3}OH_C$.

In the same way, $OG_D = \frac{1}{3}OH_D$. It follows that in the triangle OH_CH_D, the side H_CH_D is parallel to the line G_CG_D and $H_CH_D = 3G_CG_D$.

Combining these two results, we obtain that the segments CD and H_CH_D are parallel and have equal lengths. Thus, the quadrilaterals $ABCD$ and $H_AH_BH_CH_D$ have the corresponding sides parallel and of equal lengths. This proves the statement.

Second solution. Other geometric proofs can be obtained by using in various ways the Euler line. For instance, it is known that in any triangle ABC the median CM intersects the segment OH at the centroid G such that $OG = \frac{1}{3}OH$. Since CH and OM are both perpendicular to AB, it follows that the triangles CGH and MGO are similar with ratio 2:1. Hence, CH and OM are parallel and $CH = 2OM$.

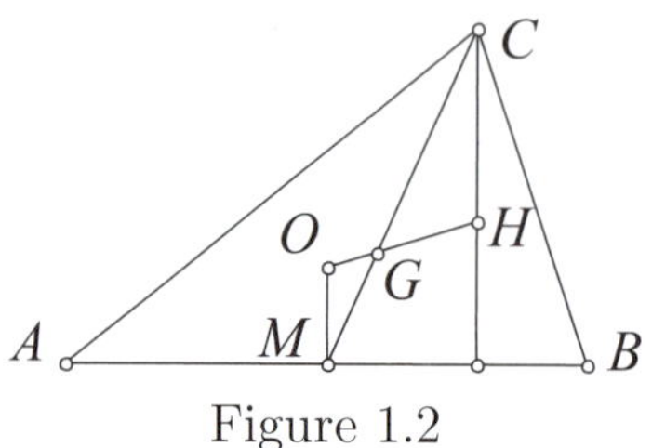

Figure 1.2

Applying the above argument to the triangles ABC and ABD, which are inscribed in the same circle, we obtain that the segments CH_D and DH_C are parallel and have the same length. Therefore, the quadrilateral CH_DH_CD is a parallelogram (see Figure 1.3). It follows that CD and H_CH_D are parallel and have equal lengths and the proof ends as the previous one.

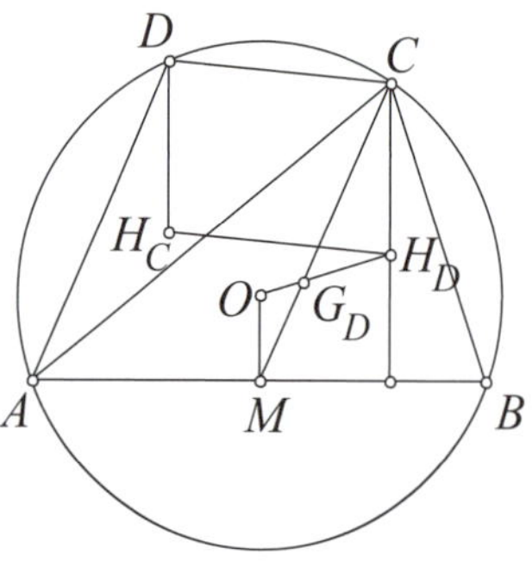

Figure 1.3

Third solution. Computational solutions are also possible, using either complex numbers, vectors or coordinates. For shortness, we will use complex numbers. Assume that the circumcenter O is the origin of the complex plane and denote by a, b, c, d the complex numbers corresponding to the points A, B, C, and D, respectively. Since O is the circumcenter of any of the triangles BCD, CDA, DAB, and ABC, the complex numbers corresponding to their orthocenters are $h_A = b + c + d$, $h_B = a + c + d$, $h_C = a + b + d$, and $h_D = a + b + c$, respectively.

Note that $h_B - h_A = a - b$, thus the vectors $\overrightarrow{H_A H_B}$ and $\overrightarrow{AB}$ are parallel, have the same length and distinct orientations. Using the same argument for the other sides of the quadrilaterals, we obtain the desired conclusion.

Observation. From the above solution one may easily obtain the following characterization of the quadrilateral $H_A H_B H_C H_D$. Let S be the point corresponding to the complex number $s = \frac{1}{2}(a + b + c + d)$. Then

$$a + h_a = b + h_b = c + h_c = d + h_d = 2s.$$

These equalities show that H_A, H_B, H_C, and H_D are the reflections of the points A, B, C, and D across the point S[2]. This, again, proves the statement of the problem (see Figure 1.4).

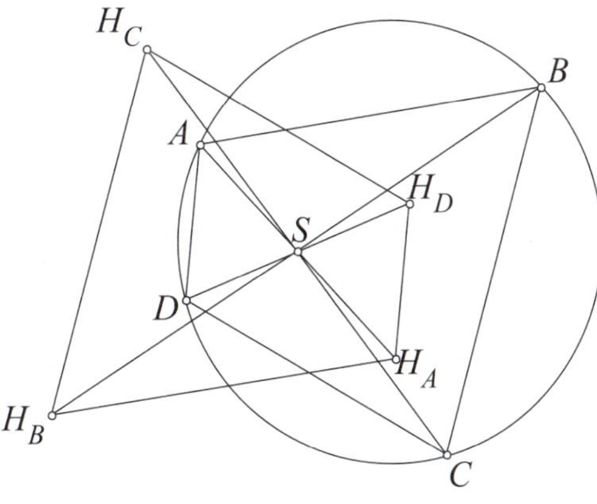

Figure 1.4

1.3. Assume that

$$5^m = \overline{a_{k-1} a_{k-2} \ldots a_1 a_0},$$

that is, 5^m has k digits. We have to find some positive integer n such that

$$5^n = \overline{a_l a_{l-1 \ldots} a_k a_{k-1} \ldots a_1 a_0}.$$

This is equivalent to: $5^n \equiv 5^m \pmod{10^k}$. The condition

$$10^k | 5^m \left(5^{n-m} - 1\right)$$

[2]The point S is called the Mathot point of the quadrilateral and it is also the point of intersection between the perpendiculars dropped from the midpoint of each side to the opposite side.

requires $k \le m$. In the same time, we have

$$10^{k-1} < 5^m < 10^k,$$

implying

$$\frac{k-1}{m} < \lg 5 < \frac{k}{m} \le 1.$$

These conditions determine k (namely, $k = \lfloor m \lg 5 \rfloor + 1$) and we have to find n such that

$$2^k | 5^{n-m} - 1.$$

There are several ways to obtain such numbers n.

The first idea is to use Euler's theorem: since $\gcd\left(5, 2^k\right) = 1$, it follows that

$$5^{\varphi\left(2^k\right)} \equiv 1 \left(\bmod 2^k\right),$$

and hence, setting $n = m + \varphi\left(2^k\right)$ yields the required result.

The second idea is to obtain the exponent n by induction. Indeed, the following statement can be easily proved: for any $s \ge 1$,

$$5^{2^s} \equiv 1 \left(\bmod 2^{s+1}\right).$$

Obviously, this is true for $s = 1$, and the induction step follows from the factorization

$$5^{2^{s+1}} - 1 = \left(5^{2^s} - 1\right)\left(5^{2^s} + 1\right).$$

Setting $n = m + 2^{k-1}$ ends the proof.

1.4. Adding up the equalities yields

$$ax + by + cz = \frac{1}{2}\left[(x-y)^2 + (y-z)^2 + (z-x)^2\right].$$

Using this equality and the given equations we obtain

$$\begin{cases} ax = (x-y)(x-z) \\ by = (y-z)(y-x) \\ cz = (z-x)(z-y) \end{cases} \tag{3}$$

Multiplying these equalities by $z-y$, $x-z$, and $y-x$, respectively, we get

$$\begin{cases} ax(z-y) = (x-y)(y-z)(z-x) \\ by(x-z) = (x-y)(y-z)(z-x) \\ cz(y-x) = (x-y)(y-z)(z-x) \end{cases} \tag{4}$$

Denote by $P = (x-y)(y-z)(z-x)$; then, from (4) it follows that

$$x(z-y) = \frac{P}{a},\ y(x-z) = \frac{P}{b},\ z(y-x) = \frac{P}{c}.$$

Adding up these equalities yields

$$P\left(\frac{1}{a}+\frac{1}{b}+\frac{1}{c}\right)=0,$$

and since a, b, and c are positive numbers, it follows that $P=0$. Hence at least two of the numbers x, y, z are equal. If, for instance, $x=y$, we derive from (3) that $x=y=0$ and $cz=z^2$, yielding the solutions $(0,0,0)$ and $(0,0,c)$. Analogously, we obtain the solutions $(a,0,0)$ and $(0,b,0)$.

Second solution. Assume that $x \leq y \leq z$. Then, from (3) we derive that $ax \geq 0$, $by \leq 0$, and $cz \geq 0$. Since a, b, c are positive numbers, we deduce that $0 \leq x \leq y \leq 0$. Hence $x=y=0$ and either $z=0$ or $z=c$. Considering the other possible orderings of the numbers x, y, z, we obtain the other solutions.

Third solution. The system (3) can be solved by brute force: denote by $u=x-z$, $v=y-z$. Then (3) is equivalent to

$$\left\{\begin{array}{l} az=u\,(u-v-a) \\ bz=v\,(v-u-b) \\ cz=uv \end{array}\right. \tag{5}$$

We analyze four cases.

Case 1. $u=v=0$. Then from (5) it follows that $x=y=z=0$, yielding the solution $(0,0,0)$.

Case 2. $u=0$ and $v \neq 0$. We immediately derive that $x=z=0$ and the second equation in (5) becomes $v-u-b=0$, hence $y=b$. We obtain thus the solution $(0,b,0)$.

Case 3. $u \neq 0$ and $v=0$. As in the previous case, we obtain the solution $(a,0,0)$.

Case 4. $u \neq 0$ and $v \neq 0$. From (5) we obtain

$$\frac{a}{c}=\frac{u-v-a}{v}$$

and

$$\frac{b}{c}=\frac{v-u-b}{u}.$$

These equalities can be translated into a system of linear equations:

$$\left\{\begin{array}{l} cu-(a+c)\,v=ac \\ (b+c)\,u-cv=-bc \end{array}\right.$$

Solving for u and v, we get $u=v=-c$, hence $z=c$ and $x=y=0$. Thus, we obtain the last solution $(0,0,c)$.

The 2nd BMO

The second Balkan Mathematical Olympiad for high-school students was held between May 6th and May 10th, 1985, in Sofia, Bulgaria. The participating countries were Bulgaria, Greece, and Romania.

Problems

2.1. In a given triangle ABC, O is its circumcenter, D is the midpoint of the side AB and E is the centroid of the triangle ACD. Show that the lines CD and OE are perpendicular if and only if $AB = AC$.

(Bulgaria)

2.2. Let $a, b, c, d \in \left[-\frac{\pi}{2}, \frac{\pi}{2}\right]$ be real numbers such that

$$\sin a + \sin b + \sin c + \sin d = 1$$

and

$$\cos 2a + \cos 2b + \cos 2c + \cos 2d \geq \frac{10}{3}.$$

Prove that $a, b, c, d \in \left[0, \frac{\pi}{6}\right]$.

(Romania)

2.3. Let S be the set of all positive integers of the form $19a + 85b$, where a, b are arbitrary positive integers. On the real axis, the points of S are colored in red and the remaining integer numbers are colored in green. Find, with proof, whether or not there exists a point A on the real axis such that any two points with integer coordinates which are symmetrical with respect to A have necessarily distinct colors.

(Greece)

2.4. There are 1985 participants to an international meeting. In any group of three participants there are at least two who speak the same language. It is known that each participant speaks at most five languages. Prove that there exist at least 200 participants who speak the same language.

(Romania)

Solutions

2.1. Suppose that $AB = AC$. Then, the altitude from A and the median CD intersect at G, the centroid of the triangle ABC (see Figure 2.1).

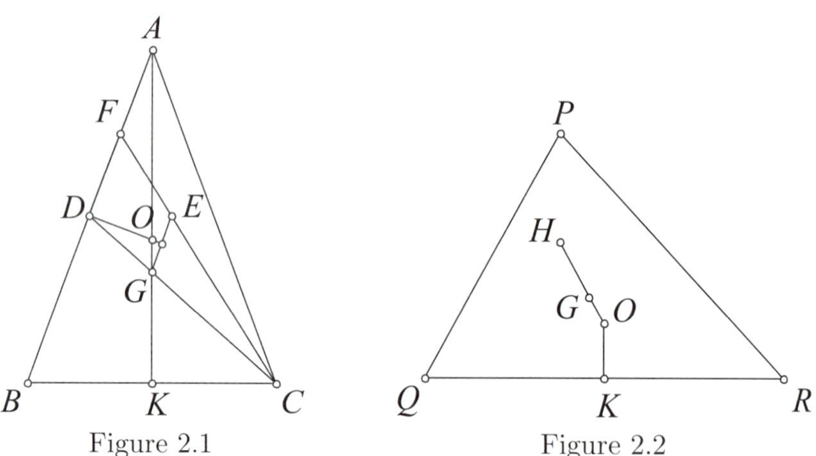

Figure 2.1 Figure 2.2

The point G lies on the segment CD and $\frac{GD}{GC} = \frac{1}{2}$. Similarly, the point E lies on the median CF of the triangle ACD and $\frac{EF}{EC} = \frac{1}{2}$. It follows that the lines AB and GE are parallel and since O lies on the perpendicular bisector of the segment AB, we conclude that DO is perpendicular to GE. Note that $\frac{DF}{DB} = \frac{EF}{EC} = \frac{1}{2}$, hence DE is parallel to BC. The point O also lies on the perpendicular bisector of the segment BC and therefore, GO is perpendicular to DE. Thus, we obtained that O is the orthocenter of the triangle GDE. Clearly, this implies that the lines CD and OE are perpendicular.

Conversely, assume that OE is perpendicular to CD. Then OE is an altitude of the triangle GDE. Like above, GE is parallel to AB, and hence the perpendicular bisector OD is perpendicular to GE. It follows that O is the orthocenter of the triangle GDE, so that GO is perpendicular to DE and BC.

Observe that in any triangle PQR, the perpendicular dropped from the circumcenter O to a side, say RQ, is the perpendicular bisector of that side (see Figure 2.2). But if the perpendicular bisector OK contains both the centroid G and the orthocenter H, then the altitude and the median from the vertex P coincide and hence $PQ = PR$.

Applying this argument to the triangle ABC yields the desired result, i.e. $AB = AC$.

Second solution. We use complex numbers: assume that the circumcenter O is the origin of the complex plane and that the triangle ABC is inscribed in the unit circle. Let a, b, c, d, e be the complex numbers corresponding to the points A, B, C, D, and E, respectively. Then

$$d = \frac{a+b}{2} \tag{1}$$

and

$$e = \frac{1}{3}\left(a + c + \frac{a+b}{2}\right) = \frac{3a+b+2c}{6}. \tag{2}$$

The lines CD and OE are orthogonal if and only if the real part of the number

$$\frac{c-d}{e}$$

equals zero, which is equivalent to

$$\frac{c-d}{e} = -\frac{\bar{c}-\bar{d}}{\bar{e}}.$$

Using (1) and (2) yields the equivalent condition

$$(2c - a - b)\left(3\bar{a} + \bar{b} + 2\bar{c}\right) = (3a + b + 2c)\left(\bar{a} + \bar{b} - 2\bar{c}\right). \tag{3}$$

Because the points A, B, and C lie on the unit circle, we have $a\bar{a} = b\bar{b} = c\bar{c} = 1$, so that (3) reduces to

$$a\bar{c} + \bar{a}c = a\bar{b} + \bar{a}b,$$

and a short computation shows that this is equivalent to

$$|a-b|^2 = |a-c|^2,$$

that is, $AB = AC$.

2.2. We denote $\sin a = x$, $\sin b = y$, $\sin c = z$, and $\sin d = u$. Thus, x, y, z, u are real numbers contained in the interval $[-1, 1]$, satisfying the condition

$$x + y + z + u = 1.$$

Moreover, using the formulas $\cos 2a = 1 - 2\sin^2 a = 1 - 2x^2$, etc. we obtain

$$x^2 + y^2 + z^2 + u^2 \leq \frac{1}{3}.$$

Since the function $\sin : \left[-\frac{\pi}{2}, \frac{\pi}{2}\right] \to [-1,1]$ is increasing, $\sin 0 = 0$ and $\sin\frac{\pi}{6} = \frac{1}{2}$, it is sufficient to prove that the numbers x, y, z, and u lie in the interval $\left[0, \frac{1}{2}\right]$.

The inequality between the arithmetic mean and the quadratic mean gives

$$\frac{x+y+z}{3} \leq \sqrt{\frac{x^2+y^2+z^2}{3}}.$$

Thus, we have

$$\frac{1}{3} \geq x^2+y^2+z^2+u^2 \geq \frac{(x+y+z)^2}{3} + u^2 = \frac{(1-u)^2}{3} + u^2,$$

which is equivalent to

$$2u^2 - u \leq 0,$$

that is, $0 \leq u \leq \frac{1}{2}$. Clearly, the same conclusion holds for x, y, z, as well.

2.3. It is well known that for any pair of integers (m, n), the following equality holds:

$$\{am + bn | a, b \in \mathbb{Z}\} = d\mathbb{Z},$$

where $d = \gcd(m, n)$ and $d\mathbb{Z} = \{dk | k \in \mathbb{Z}\}$.
In our case, we have $\gcd(19, 85) = 1$, hence

$$\{19a + 85b | a, b \in \mathbb{Z}\} = \mathbb{Z}.$$

Since a and b are positive integers, the set S will contain only positive integers. Moreover, for small values of a and b, S has some gaps with respect to the set $\mathbb{N}$ of positive integers. For instance, the least element of S is $104 = 19 + 85$, the next is $123 = 2 \cdot 19 + 85$, then $142 = 3 \cdot 19 + 85$, etc. Thus, the elements of S are

$$104 < 123 < 142 < 161 < 180 < 189 < 208 < \ldots.$$

We expect that for large numbers, the density of S will increase. Indeed, this is a consequence of the following result:
Lemma. Let $m, n > 1$ be coprime positive integers and let

$$S = \{am + bn | a, b \in \mathbb{Z}, a \geq 1, b \geq 1\}.$$

Then $mn \notin S$ and S contains all integers N satisfying the condition $N > mn$.
Proof. For the first statement, let us assume, by way of contradiction, that $mn \in S$, and let $a, b > 0$ such that $mn = am + bn$. It is obvious that $a < n$ and $b < m$. Since $m(n-a) = bn$ and $\gcd(m, n) = 1$, we obtain that m divides b. This contradicts the condition $0 < b < m$.

For the second statement, let $N > mn$ be an arbitrary integer. Because $\gcd(m, n) = 1$, the congruence

$$N \equiv mx \pmod{n}$$

has a solution $x = a$, where $0 < a \leq n$. But $N > mn \geq am$, and hence $N - am = bn$, for some positive integer b. The lemma is proved.

Returning to our problem, we will prove that a point A with the required property indeed exists. We consider the general case, replacing 19 and 85 by two coprime integers m, $n > 1$.

Let a be the coordinate of the point A. Note that A is the midpoint of a segment whose endpoints have coordinates x and y if and only if $x + y = 2a$.

Because $mn \notin S$, the integer x such that $mn + x = 2a$ will necessary belong to S, and therefore $x \geq m + n$. We derive the condition

$$a \geq \frac{mn + m + n}{2}.$$

The integer y such that $m + n + y = 2a$ is not in S, because $m + n \in S$. Therefore $y \leq mn$, and we deduce that

$$a \geq \frac{mn + m + n}{2}.$$

Consequently, if such a point A exists, its coordinate must be

$$a = \frac{mn + m + n}{2}.$$

We now prove that this point indeed satisfies the required condition. Let $x < y$ be two integers such that

$$x + y = 2a = mn + m + n.$$

If $x = m + n$, then $y = mn$ and we have $x \in S$, $y \notin S$.

If $x < m + n$, then $y > mn$. Clearly, $x \notin S$ (note that $m + n$ is the least element of S) and $y \in S$, according to the lemma.

Finally, suppose that $m+n < x < a < y < mn$. We can prove that x and y cannot both belong to S. Indeed, suppose that $x = am+bn$ and $y = a'm+b'n$, for some positive integers a, b, a', and b'. Then, the equality $x+y = mn+m+n$ yields

$$\left(a + a' - 1\right) m + \left(b + b' - 1\right) n = mn.$$

Since $a + a' - 1$ and $b + b' - 1$ are positive integers, it follows that $mn \in S$, a contradiction.

To end the proof, it is sufficient to prove that at least one of the numbers x and y belongs to S. We distinguish two cases:

Case 1. The number x is not divisible by m or by n. In this case, the congruence

$$x \equiv bn \pmod{m}$$

has a solution b, with $0 < b < m$, and hence $x - bn = am$, for some $a > 0$. It follows that $x = am + bn \in S$.

Case 2. The number x is divisible by either m or n. Assume, for instance, that $x = am$, for some positive integer a. Then $y = mn + m + n - am = (n + a + 1)\,m + n \in S$, as desired.

For $m = 19$ and $n = 85$, the required point A has the coordinate

$$a = \frac{19 \cdot 85 + 19 + 85}{2} = \frac{1719}{2}.$$

We add that in all cases, a is a noninteger rational number, since $mn+m+n$ is an odd integer.

2.4. Let A be one of the participants to the meeting. We know that he speaks at most five languages. Assume that every other participant speaks at least one of these languages. Then at least $\left\lfloor \frac{1984}{5} \right\rfloor + 1 = 397$ participants speak the same language and we are done. Otherwise, there exists a participant B who speaks other five languages (at most). This gives a total of at most ten languages spoken by A and B together. But every other participant speaks at least one of these ten languages, since in any group of three there exist at least two speaking the same language. Therefore, at least $\left\lfloor \frac{1983}{10} \right\rfloor + 1 = 199$ participants speak a common language with either A or B, which gives a total of 200 participants speaking the same language.

The 3rd BMO

The third Balkan Mathematical Olympiad for high-school students was held between May 1st and May 5th , 1986, in Bucharest, Romania. The participating countries were Bulgaria, Greece, and Romania.

Problems

3.1. A line passing through the incenter I of the triangle ABC intersects its incircle at D and E and its circumcircle at F and G, in such a way that the point D lies between I and F. Let r be the inradius of the triangle ABC. Prove that

$$DF \cdot FG \geq r^2.$$

When does the equality hold?

(Greece)

3.2. Let $ABCD$ be a tetrahedron and let E, F, G, H, K, L be points lying on the edges AB, BC, CD, DA, DB, and DC, respectively, in such a way that

$$AE \cdot BE = BF \cdot CF = CG \cdot AG = DH \cdot AH = DK \cdot BK = DL \cdot CL.$$

Prove that the points E, F, G, H, I, J, K, L lie on a sphere.

(Bulgaria)

3.3. Let a, b, c be real numbers such that $ab \neq 0$, $c > 0$, and let $(a_n)_{n \geq 1}$ be the sequence of real numbers defined by: $a_1 = a$, $a_2 = b$, and

$$a_{n+1} = \frac{a_n^2 + c}{a_{n-1}}$$

for all $n \geq 2$.

Show that the terms of the sequence are all integer numbers if and only if the numbers a, b, and $\dfrac{a^2 + b^2 + c}{ab}$ are integers.

(Bulgaria)

3.4. Let ABC be a triangle and let P be a point in its plane such that the triangles PAB, PBC and PCA have equal areas and equal perimeters. Prove that:

a) if P lies in the interior of the triangle ABC, then the triangle is equilateral;

b) if P lies in the exterior of the triangle ABC, then the triangle is right angled.

(Romania)

Solutions

3.1. Denote by R the circumradius of the triangle ABC. It is not difficult to see that

$$DF \cdot EG = (IF - r)(IG - r) = IF \cdot IG - r(IF + IG) + r^2.$$

Therefore, $DF \cdot EG \geq r^2$ if and only if $IF \cdot IG \geq r \cdot FG$.

The power of a point theorem, applied to the point I gives

$$IF \cdot IG = R^2 - OI^2.$$

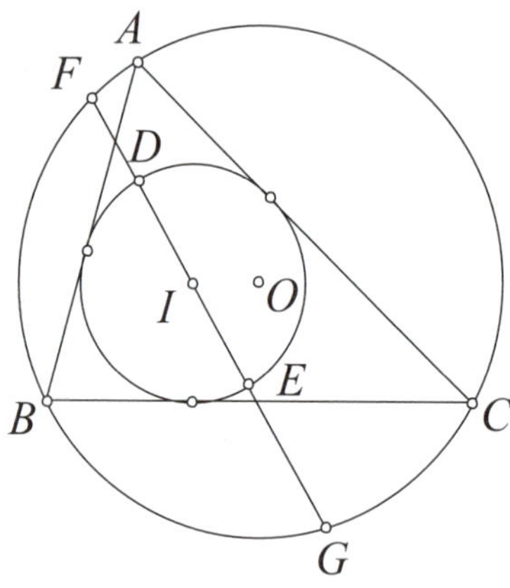

Figure 3.1

On the other hand, the Euler triangle formula states that

$$R^2 - OI^2 = 2Rr.$$

Consequently, we have to prove that

$$2Rr \geq r \cdot FG,$$

or

$$2R \geq FG,$$

which is obvious, since FG is a chord in the circumcircle of the triangle ABC.

The equality occurs if and only if $FG = 2R$, that is, the line FG passes through both I and O. If the triangle ABC is equilateral, then any diameter satisfy the condition, since in this case, I and O coincide. If ABC is not equilateral, then the equality holds if and only if the line FG coincide with OI.

3.2. Let $\mathcal{S}$ be the tetrahedron's circumscribed sphere, R its radius and O its center. Let π be the plane determined by the points A, B, and O. The intersection between π and $\mathcal{S}$ is a circle $\mathcal{C}$ of radius R. We apply the power of a point theorem to the point E lying inside the circle $\mathcal{C}$. This gives

$$AE \cdot BE = R^2 - OE^2.$$

In a similar way, we obtain $BF \cdot CF = R^2 - OF^2$, $CG \cdot AG = R^2 - OG^2$, etc. Clearly, this implies

$$OE = OF = OG = OH = OK = OL,$$

and hence, the points E, F, G, H, K, and L lie on a sphere centered at O.

3.3. Note that $a_n \neq 0$, for all $n \geq 1$. The recursive relation can be written under the form

$$a_{n+1}a_{n-1} - a_n^2 = c,$$

for all $n \geq 2$. This implies

$$a_{n+1}a_{n-1} - a_n^2 = a_{n+2}a_n - a_{n+1}^2,$$

and hence,

$$\frac{a_{n+2} + a_n}{a_{n+1}} = \frac{a_{n+1} + a_{n-1}}{a_n},$$

for all $n \geq 2$. Iterating the latter gives

$$\frac{a_{n+1} + a_{n-1}}{a_n} = \frac{a_n + a_{n-2}}{a_{n-1}} = \ldots = \frac{a_3 + a_1}{a_2},$$

and we conclude that

$$\frac{a_{n+1} + a_{n-1}}{a_n} = \frac{a^2 + b^2 + c}{ab}, \tag{1}$$

for all $n \geq 2$.

If we denote by $k = \frac{a^2+b^2+c}{ab}$, we see that (1) can be written as follows:

$$a_{n+1} = ka_n - a_{n-1}, \tag{2}$$

for all $n \geq 2$.

Suppose that a, b, and k are integer numbers. Then (2) implies that a_n is an integer number for all $n \geq 3$.

Conversely, suppose that a_n is an integer, for all $n \geq 1$. Then a and b are integers and $k = \frac{a^2+b^2+c}{ab}$ is a rational number. Let $k = \frac{p}{q}$, where p and q are coprime integers. Substituting in (2) yields

$$pa_n = q\left(a_{n+1} + a_{n-1}\right), \tag{3}$$

for all $n \geq 2$. Since $(p, q) = 1$, it follows that q divides a_n for all $n \geq 2$. Thus, for $n \geq 3$, q divides both a_{n-1} and a_{n+1}, so that we deduce, using (3), that q^2 divides a_n. A simple inductive argument shows that q^m divides a_n, for all $n \geq m+1$. But then, for sufficiently large n, the equality

$$a_{n+1}a_{n-1} - a_n^2 = c$$

implies that q^m divides c. Since m is an arbitrary positive integer, this is possible only if $q = 1$, and hence $k = \frac{a^2+b^2+c}{ab}$ is also an integer number.

Observation. The relation (1) can also be obtained in the following way:

$$\begin{aligned}\frac{a_{n+1}+a_{n-1}}{a_n} &= \frac{a_{n+1}^2 + a_{n+1}a_{n-1}}{a_n a_{n+1}} = \frac{a_n^2 + a_{n+1}^2 + c}{a_n a_{n+1}} = \frac{a_n + \left(\frac{a_n^2+c}{a_{n-1}}\right)^2 + c}{a_n \frac{a_n^2+c}{a_{n-1}}} \\ &= \frac{a_{n-1}^2 + a_n^2 + c}{a_{n-1}a_n} = \dots = \frac{a_1^2 + a_2^2 + c}{a_1 a_2} = \frac{a^2+b^2+c}{ab}.\end{aligned}$$

3.4. We first consider the case when P lies in the interior of the triangle ABC. Suppose that the lines PA, PB, and PC intersect the sides BC, CA, AB at A', B', C', respectively. We have

$$\frac{A'B}{A'C} = \frac{[ABA']}{[ACA']} = \frac{[PBA']}{[PCA']} = \frac{[ABA'] - [PBA']}{[ACA'] - [PCA']} = \frac{[PAB]}{[PAC]}.$$

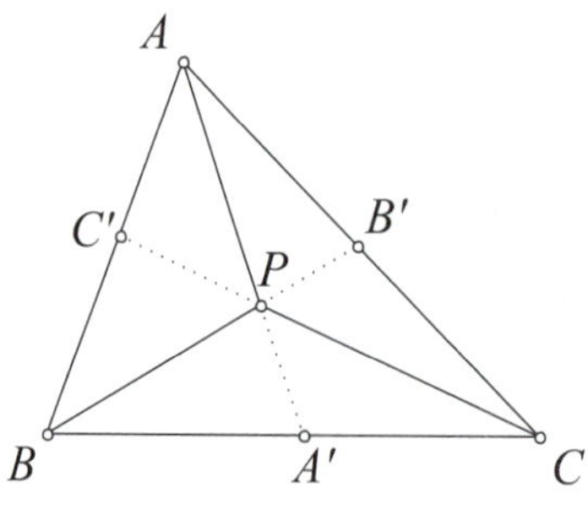

Figure 3.2

Thus, if $[PAB] = [PAC]$, then $A'B = A'C$ and P lies on the median AA'. We conclude that since the triangles PAB, PBC and PCA have equal areas, the point P is the centroid of the triangle ABC.

We will reach the required conclusion by proving the following result:
Lemma. Let G be the centroid of the triangle ABC. The triangles GAB, GBC, and GBA have equal perimeters if and only if the triangle ABC is equilateral.
Proof. All is clear if ABC is equilateral.

Conversely, suppose that the triangles GAB, GBC, and GBA have equal perimeters, and assume, by way of contradiction, that $AB < AC$. Then both points A and G lie in the halfplane defined by the perpendicular bisector of the segment BC which contains the point B (see Figure 3.3).

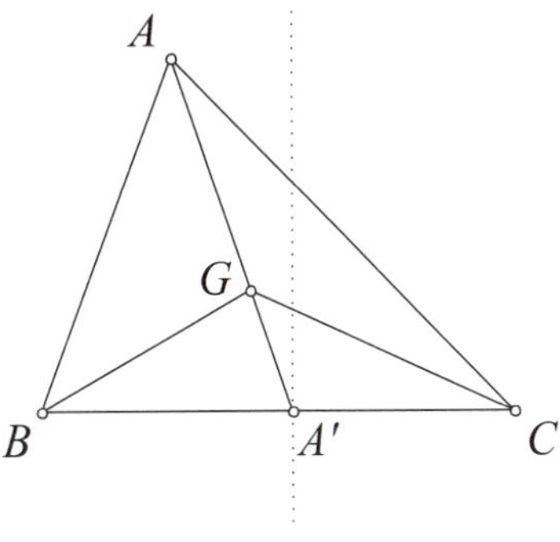

Figure 3.3

This implies that $GB < GC$. It follows that $AB+GB+GA < AC+GC+GA$, thus, the perimeters of the triangles GAB and GAC are not equal, which is a contradiction. This proves the lemma and the first part of the problem.

Next, suppose that P lies in the exterior of the triangle ABC. It is not difficult to see that P must lie in the interior of one of the regions I, II or III (see Figure 3.4), otherwise the areas of triangles PAB, PBC, and PCA cannot be equal.

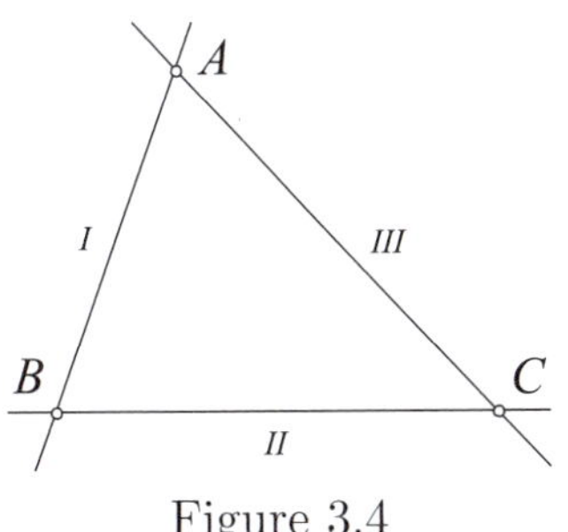

Figure 3.4

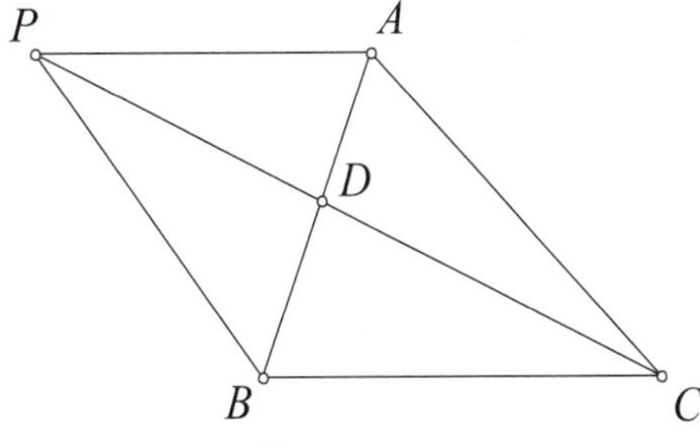

Figure 3.5

Suppose that P lies inside region I. Because $[PAB] = [PCB]$, the points A and C are equally distanced to PB, and hence AC is parallel to PB. Similarly, BC is parallel to PA, so $PACB$ is a parallelogram. Since the triangles PAB, PBC, and PCA have the equal perimeters, it follows that $PC = AB$ and therefore, $PACB$ is a rectangle. Thus, $\angle C = 90^\circ$ and the claim is proved.

The 4th BMO

The fourth Balkan Mathematical Olympiad for high-school students was held between May 3rd and May 8th, 1986, in Athens, Greece. The number of the participating countries increased to 5: Bulgaria, Cyprus, Greece, Romania and Yugoslavia.

Problems

4.1. Let a be a real number and let $f:\mathbb{R}\to\mathbb{R}$ be a function with the properties: $f(0)=\frac{1}{2}$, and

$$f(x+y)=f(x)f(a-y)+f(y)f(a-x),$$

for all real numbers x and y.

Prove that f is a constant function.

(Yugoslavia)

4.2. Find all real numbers x, y greater than 1, satisfying the condition that the numbers $\sqrt{x-1}+\sqrt{y-1}$ and $\sqrt{x+1}+\sqrt{y+1}$ are nonconsecutive integers.

(Romania)

4.3. In the triangle ABC the following equality holds:

$$\sin^{23}\frac{A}{2}\cos^{48}\frac{B}{2}=\sin^{23}\frac{B}{2}\cos^{48}\frac{A}{2}.$$

Determine the value of $\dfrac{AC}{BC}$.

(Cyprus)

4.4. The circles K_1 and K_2, centered at O_1 and O_2, with radii 1 and $\sqrt{2}$, respectively, intersect at A and B. Let C be a point on K_2 such that the midpoint of AC lies on K_1. Find the length of AC if $O_1O_2=2$.

(Bulgaria)

Solutions

4.1. Plugging $x = y = 0$ in the given relation yields

$$f(0) = 2f(0)f(a),$$

hence $f(a) = \frac{1}{2}$. For $y = 0$ we obtain

$$f(x) = f(x)f(a) + f(0)f(a-x),$$

which implies

$$f(x) = f(a-x),$$

for all real x.

Setting $y = a - x$ in the initial relation gives

$$f(a) = f^2(x) + f^2(a-x),$$

and thus

$$f^2(x) = \frac{1}{4},$$

for all real x.

Replacing both x and y in the initial relation with $\frac{x}{2}$ yields

$$f(x) = 2f\left(\frac{x}{2}\right)f\left(a - \frac{x}{2}\right) = 2f^2\left(\frac{x}{2}\right) = 2 \cdot \frac{1}{4} = \frac{1}{2},$$

thus f is a constant function.

Second solution. As before, we obtain $f(a) = \frac{1}{2}$ and $f(x) = f(a-x)$, for all real x. Then the initial relation becomes

$$f(x+y) = 2f(x)f(y),$$

for all x and y. Replacing y by $a - y$ yields

$$f(x+a-y) = 2f(x)f(a-y) = 2f(x)f(y),$$

and hence

$$f(x+y) = f(x+a-y).$$

Finally, replace x and y with $\frac{x}{2}$. We obtain

$$f(x) = f(a) = \frac{1}{2},$$

for all real x, and we are done.

4.2. Let $A = \sqrt{x-1} + \sqrt{y-1}$ and $B = \sqrt{x+1} + \sqrt{y+1}$. Then

$$\begin{aligned} B - A &= \sqrt{x+1} - \sqrt{x-1} + \sqrt{y+1} - \sqrt{y-1} \\ &= \frac{2}{\sqrt{x+1} + \sqrt{x-1}} + \frac{2}{\sqrt{y+1} + \sqrt{y-1}}. \end{aligned}$$

Because $x, y > 1$, we have $\sqrt{x+1} + \sqrt{x-1} > \sqrt{2}$ and $\sqrt{y+1} + \sqrt{y-1} > \sqrt{2}$. It follows that $B - A < 2\sqrt{2} < 3$, and since A and B are nonconsecutive integers we conclude that $B - A = 2$.

Suppose with no loss of generality that $x \le y$. Then the equality

$$\frac{2}{\sqrt{x+1} + \sqrt{x-1}} + \frac{2}{\sqrt{y+1} + \sqrt{y-1}} = 2$$

implies $x \le \frac{5}{4} \le y$. Indeed, note that $\sqrt{\frac{5}{4} + 1} + \sqrt{\frac{5}{4} - 1} = 2$, thus x and y cannot be both greater or both less than $\frac{5}{4}$. Moreover, since $x \ge 1$, we have

$$\frac{2}{\sqrt{x+1} + \sqrt{x-1}} \le \frac{2}{\sqrt{2}},$$

and then

$$2 - \frac{2}{\sqrt{y+1} + \sqrt{y-1}} \le \sqrt{2}.$$

The latter is equivalent to

$$\sqrt{y+1} + \sqrt{y-1} \le 2 + \sqrt{2},$$

and hence $y \le 3$.

Consequently, we have

$$A = \sqrt{x-1} + \sqrt{y-1} \ge \frac{1}{2}$$

and

$$B = \sqrt{x+1} + \sqrt{y+1} \le \frac{3}{2} + 2,$$

thus $A = 1$ and $B = 3$.

If we denote by $a = \sqrt{x-1} + \sqrt{x+1}$ and $b = \sqrt{y-1} + \sqrt{y+1}$, we obtain

$$\begin{aligned} a + b &= 4 \\ \frac{1}{a} + \frac{1}{b} &= 1, \end{aligned}$$

hence a and b are the roots of the equation $t^2 - 4t + 4 = 0$. This implies $a = b = 2$ and, finally, $x = y = \frac{5}{4}$.

4.3. We claim that $\frac{AC}{BC} = 1$, that is, the triangle ABC is isosceles. Suppose, by way of contradiction, that $AC > BC$. Then $A < B$, and therefore

$$0 < \frac{A}{2} < \frac{B}{2} < \frac{\pi}{2}.$$

Because the function $f(x) = \sin x$ is increasing on the interval $\left(0, \frac{\pi}{2}\right)$, we have

$$0 < \sin\frac{A}{2} < \sin\frac{B}{2},$$

and hence

$$0 < \sin^{23}\frac{A}{2} < \sin^{23}\frac{B}{2}.$$

The function $g(x) = \cos x$ is decreasing on the same interval, therefore

$$0 < \cos\frac{B}{2} < \cos\frac{A}{2},$$

so that

$$0 < \cos^{48}\frac{B}{2} < \cos^{48}\frac{A}{2}.$$

Multiplying these inequalities yields

$$\sin^{23}\frac{A}{2}\cos^{48}\frac{B}{2} < \sin^{23}\frac{B}{2}\cos^{48}\frac{A}{2},$$

which contradicts the hypothesis.

In a similar way, if we assume that $AC < BC$, we again reach a contradiction.

4.4. Let M be the midpoint of the line segment AC and let N be a point on K_1 such that AN is a diameter. It is not difficult to see that M lies on a circle which is the image of K_2 through a homothety of center A and ratio $\frac{1}{2}$ (see Figure 4.1).

Clearly, $\angle AMN = \angle AMO_2 = 90°$, hence the points O_2, M, and N are collinear and AM is an altitude in the triangle AO_2N. In the same triangle, O_2O_1 is a median. It follows that

$$O_1O_2^2 = \frac{2\left(AO_2^2 + NO_2^2\right) - AN^2}{4}.$$

But $O_1O_2 = 2$, $AO_2 = \sqrt{2}$, and $AN = 2$. Replacing these values in the above equality gives $NO_2 = 2\sqrt{2}$.

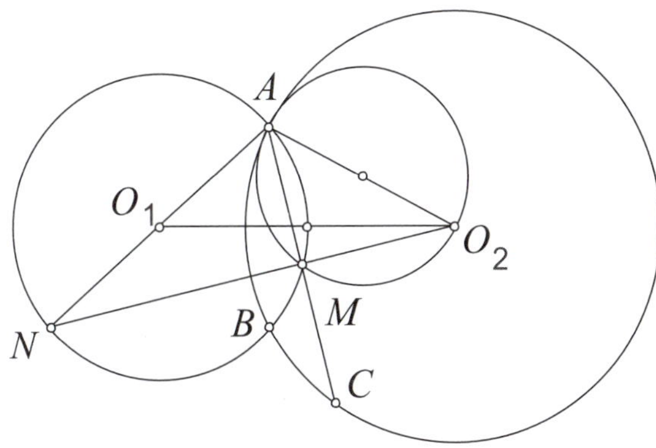

Figure 4.1

We will determine the length of the line segment AM expressing in two ways the area of the triangle ANO_2. We have

$$[ANO_2] = \frac{1}{2}AM \cdot NO_2 = AM\sqrt{2}.$$

On the other hand, Heron's formula gives

$$[ANO_2] = \frac{\sqrt{7}}{2}.$$

We deduce that $AM = \frac{\sqrt{14}}{4}$ and hence $AC = \frac{\sqrt{14}}{2}$.
Observation. It is often useful to write Heron's formula under the form

$$16\,[ABC]^2 = 2\left(a^2b^2 + b^2c^2 + c^2a^2\right) - a^4 - b^4 - c^4,$$

where a, b, c denote the side lengths of the triangle ABC.

Second solution. We use coordinates. Let K_1 be the unit circle and O_1O_2 be the Ox axis (see Figure 4.2). We denote by (x_P, y_P) the coordinates of an arbitrary point P.

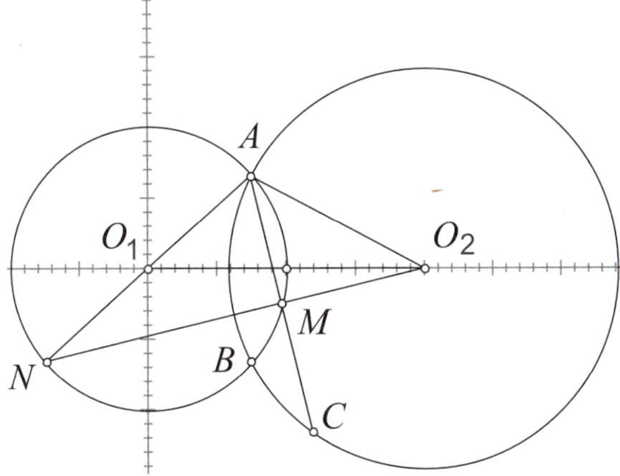

Figure 4.2

We have $O_1A = 1$ and $O_2A = \sqrt{2}$, so that the coordinates of the point A satisfy the system of equations

$$\begin{cases} x_A^2 + y_A^2 = 1 \\ (x_A - 2)^2 + y_A^2 = 2. \end{cases}$$

Solving for x_A and y_A we obtain $x_A = \frac{3}{4}$ and $y_A = \frac{\sqrt{7}}{4}$ (we assumed that $y_A > 0$, that is, the point A lies above the Ox axis, as in Figure 4.2).

Since M is the midpoint of AC, we have

$$x_M = \frac{x_A + x_C}{2}, \; y_M = \frac{y_A + y_C}{2},$$

hence

$$x_C = 2x_M - \frac{3}{4}, \; y_C = 2y_M - \frac{\sqrt{7}}{4}.$$

But $O_2C = \sqrt{2}$, therefore

$$(x_C - 2)^2 + y_C^2 = 2,$$

and it follows that

$$\left(2x_M - \frac{11}{4}\right)^2 + \left(2y_M - \frac{\sqrt{7}}{4}\right)^2 = 2.$$

Since M lies on K_1, we obtain the system

$$\begin{cases} x_M^2 + y_M^2 = 1 \\ \left(2x_M - \frac{11}{4}\right)^2 + \left(2y_M - \frac{\sqrt{7}}{4}\right)^2 = 2, \end{cases}$$

whose solutions are $\left(\frac{3}{4}, \frac{\sqrt{7}}{4}\right)$ – the coordinates of A – and $\left(\frac{31}{32}, -\frac{3\sqrt{7}}{32}\right)$.

Finally, we obtain $x_C = \frac{19}{16}$ and $y_C = -\frac{7}{16}\sqrt{7}$, hence

$$AC = \sqrt{\left(\frac{3}{4} - \frac{19}{16}\right)^2 + \left(\frac{\sqrt{7}}{4} + \frac{7\sqrt{7}}{16}\right)^2} = \frac{\sqrt{14}}{2}.$$

The 5th BMO

The fifth Balkan Mathematical Olympiad for high-school students was held between May 1st and May 7th , 1987, in Nicosia, Cyprus. The number of the participating countries remained 5: Bulgaria, Cyprus, Greece, Romania and Yugoslavia.

Problems

5.1. Let ABC be a triangle and let M, N, and P be points on the line BC such that AM, AN, and AP are the altitude, the angle bisector and the median of the triangle, respectively. It is known that

$$\frac{[AMP]}{[ABC]} = \frac{1}{4} \text{ and } \frac{[ANP]}{[ABC]} = 1 - \frac{\sqrt{3}}{2}.$$

Find the angles of the triangle ABC.

(Bulgaria)

5.2. Find all polynomials of two variables $P(x, y)$ which satisfy

$$P(a,b)\,P(c,d) = P(ac+bd, ad+bc),$$

for all real numbers a, b, c, d.

(Yugoslavia)

5.3. Let $ABCD$ be a tetrahedron and let d be the sum of squares of its edges' lengths. Prove that the tetrahedron can be included in a region bounded by two parallel planes, the distance between the planes being at most $\dfrac{\sqrt{d}}{2\sqrt{3}}$.

(Greece)

5.4. Let $(a_n)_{n\geq 1}$ be a sequence defined by $a_n = 2^n + 49$. Find all values of n such that $a_n = pq$, $a_{n+1} = rs$, where p, q, r, s are prime numbers with $p < q$, $r < s$ and $q - p = s - r$.

(Romania)

Solutions

5.1. It is not difficult to see that the triangle ABC is not isosceles. Suppose that $AB < AC$; then M lies on the halfline $(PB$. Because P is the midpoint of the line segment BC, we have

$$\frac{[APB]}{[ABC]} = \frac{1}{2}$$

and therefore

$$\frac{[AMP]}{[APB]} = \frac{1}{2}.$$

This shows that M lies inside the line segment BP and that it is its midpoint (see Figure 5.1). It follows that APB is an isosceles triangle and $AB = AP$.

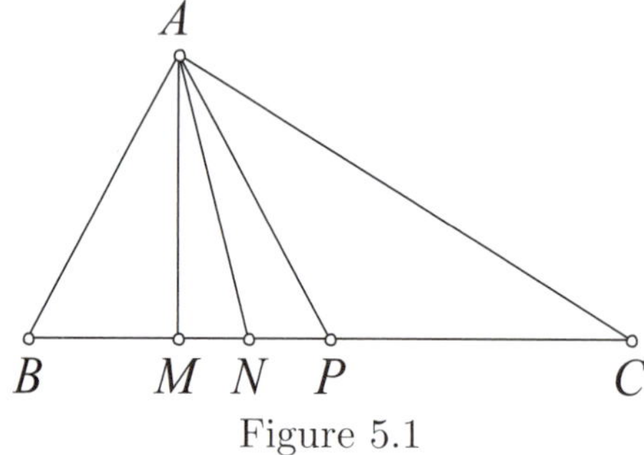

Figure 5.1

On the other hand, we have

$$\frac{[ANP]}{[ABC]} = \frac{NP}{BC} = 1 - \frac{\sqrt{3}}{2}.$$

The internal angle bisector theorem yields

$$\frac{AB}{AC} = \frac{NB}{NC},$$

hence

$$\frac{AB}{AC} = \frac{BP - NP}{CP + NP} = \frac{\frac{BP-NP}{BC}}{\frac{CP+NP}{BC}} = \frac{\frac{1}{2} - \left(1 - \frac{\sqrt{3}}{2}\right)}{\frac{1}{2} + \left(1 - \frac{\sqrt{3}}{2}\right)} = \frac{1}{\sqrt{3}}.$$

Now,

$$\frac{1}{9} = \left(\frac{BM}{MC}\right)^2 = \frac{AB^2 - AM^2}{AC^2 - AM^2} = \frac{1 - \left(\frac{AM}{AB}\right)^2}{\left(\frac{AC}{AB}\right)^2 - \left(\frac{AM}{AB}\right)^2} = \frac{1 - \sin^2 B}{3 - \sin^2 B},$$

and a short computation shows that

$$\sin \angle B = \frac{\sqrt{3}}{2},$$

hence $\angle B = 60^\circ$. We deduce that the triangle ABP is equilateral so it follows that $AP = PC$. Then the angle $\angle APB$ (which is an exterior angle of the triangle APC) is twice the angle $\angle C$.

Finally, we obtain $\angle C = 30^\circ$ and $\angle A = 90^\circ$.

5.2. The relation from the hypothesis gives

$$P(x,0)\,P(x,0) = P\left(x^2,0\right),$$

for all real x. Thus, the polynomial $Q(x) = P(x,0)$ satisfies

$$Q^2(x) = Q\left(x^2\right), \tag{1}$$

for all real (and hence also for all complex) numbers x. If Q is a constant polynomial it follows easily that either $Q = 0$ or $Q = 1$. If Q is nonconstant, we claim that

$$Q(x) = x^n,$$

for some positive integer n. Indeed, condition (1) implies that if α is a root of Q, then β, with the property that $\beta^2 = \alpha$, is also a root of Q. Suppose that α is a nonzero (complex) root of Q. Then we can write

$$\alpha = r(\cos\theta + i\sin\theta),$$

for some $r > 0$ and $\theta \in [0, 2\pi)$. Using de Moivre's formula, we deduce that the numbers $r^{\frac{1}{2}}\left(\cos\frac{\theta}{2} + i\sin\frac{\theta}{2}\right), r^{\frac{1}{4}}\left(\cos\frac{\theta}{4} + i\sin\frac{\theta}{4}\right), \ldots$ are also roots of Q, which is impossible, since Q cannot have infinitely many distinct roots. It follows that all the roots of Q are equal to zero. Using (1) again, it is easy to see that the leading coefficient of Q equals 1, that is, $Q(x) = x^n$, for some integer n.

If $Q(x) = 0$, then

$$P(x,y) = P(1,0)\,P(x,y) = Q(1)\,P(x,y) = 0,$$

for all x and y.

If $Q(x) = x^n$, for some $n \geq 0$, then observe that

$$P(tx,ty) = P(t,0)\,P(x,y) = t^n P(x,y),$$

hence P is a homogenous polynomial of degree n, that is,

$$P(x,y) = a_0x^n + a_1x^{n-1}y + a_2x^{n-2}y^2 + \ldots + a_{n-1}xy^{n-1} + a_ny^n.$$

Let

$$f(x) = a_0 x^n + a_1 x^{n-1} + a_2 x^{n-2} + \ldots + a_{n-1} x + a_n.$$

Then

$$P(x,y) = y^n f\left(\frac{x}{y}\right),$$

for all x, y with $y \neq 0$. The hypothesis yields

$$b^n f\left(\frac{a}{b}\right) d^n f\left(\frac{c}{d}\right) = (ad+bc)^n f\left(\frac{ac+bd}{ad+bc}\right) = b^n d^n \left(\frac{a}{b}+\frac{c}{d}\right)^n f\left(\frac{\frac{a}{b}\frac{c}{d}+1}{\frac{a}{b}+\frac{c}{d}}\right),$$

or

$$f(x) f(y) = (x+y)^n f\left(\frac{xy+1}{x+y}\right), \tag{2}$$

for all x, y with $x + y \neq 0$.

We claim that the only possible roots of f are -1 and 1. Indeed, suppose that $y_0 \neq \pm 1$ is a root, and let $z_0 \neq y_0$ be an arbitrary number. Setting

$$x_0 = \frac{y_0 z_0 - 1}{y_0 - z_0},$$

the above equality gives

$$0 = f(x_0) f(y_0) = (x_0 + y_0)^n f(z_0).$$

We cannot have $x_0 + y_0 = 0$, since this would lead to

$$\frac{y_0 z_0 - 1}{y_0 - z_0} + y_0 = \frac{y_0^2 - 1}{y_0 - z_0} = 0,$$

which is impossible. It follows that $f(z_0) = 0$, and hence f has infinitely many roots, a contradiction.

Finally, we derive that $f(x) = a(x-1)^k (x+1)^{n-k}$ for some integer k, $0 \leq k \leq n$. Plugging this in (2) gives $a = 1$ and then

$$P(x,y) = y^n f\left(\frac{x}{y}\right) = (x-y)^k (x+y)^{n-k}.$$

Second solution. Define the polynomial Q as

$$Q(x,y) = P\left(\frac{x+y}{2}, \frac{x-y}{2}\right).$$

It is not difficult to see that P can be expressed in terms of Q as follows:

$$P(x,y) = Q(x+y, x-y).$$

We claim that Q satisfies the condition

$$Q(x,y)Q(z,t) = Q(xz,yt), \tag{3}$$

for all real x, y, z, t. Indeed, since

$$P(a,b)P(c,d) = P(ac+bd, ad+bc),$$

we have

$$\begin{aligned} Q(a+b,a-b)Q(c+d,c-d) &= Q(ac+bd+ad+bc, ac+bd-ad-bc) \\ &= Q((a+b)(c+d),(a-b)(c-d)). \end{aligned}$$

Therefore, denoting $x = a+b$, $y = a-b$, $z = c+d$, $t = c-d$, we obtain (3). Clearly, since a, b, c, d are arbitrary real numbers, x, y, z, t can take any real values as well.

Now, the polynomials

$$Q_1(x) = Q(x,1), \; Q_2(x) = Q(1,x)$$

satisfy

$$\begin{aligned} Q_1(xy) &= Q_1(x)Q_1(y), \\ Q_2(xy) &= Q_2(x)Q_2(y), \end{aligned}$$

and

$$Q(x,y) = Q_1(x)Q_2(y),$$

for all real x, y. Using the technique displayed in the previous solution we deduce that $Q_1(x) = 0$, or $Q_2(x) = 0$, or

$$Q_1(x) = x^k, \; Q_2(x) = x^l,$$

for some integers $k, l \geq 0$. It follows that either $Q(x,y) = 0$ and hence $P(x,y) = 0$, or

$$Q(x,y) = x^k x^l,$$

and then

$$P(x,y) = (x+y)^k (x-y)^l.$$

5.3. Let M, N, P, Q, R, and S be the midpoints of the edges AB, BC, CD, DA, AC, and BD, respectively (see Figure 5.2). In the triangle ABD, MQ is parallel to BD and $MQ = \frac{1}{2}BD$. Similarly, PN is parallel to BD and $PN = \frac{1}{2}BD$. We obtain that $MNPQ$ is a parallelogram and then

$$MP^2 + NQ^2 = 2\left(MQ^2 + MN^2\right) = \frac{1}{2}\left(BD^2 + AC^2\right).$$

Indeed, the cosine law gives

$$MP^2 = MQ^2 + PQ^2 - 2MQ \cdot PQ \cos \angle MQP,$$
$$NQ^2 = MQ^2 + MN^2 - 2MQ \cdot MN \cos \angle QMN.$$

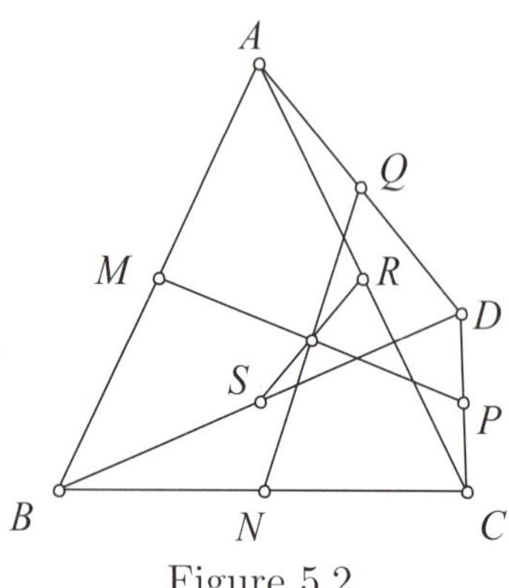

Figure 5.2

Since $PQ = MN$ and $\cos \angle MQP = -\cos \angle QMN$, the claim follows from adding up the previous equalities.

Analogously, we obtain

$$MP^2 + RS^2 = \frac{1}{2}\left(AD^2 + BC^2\right)$$

and

$$NQ^2 + RS^2 = \frac{1}{2}\left(AB^2 + CD^2\right).$$

It follows that

$$MP^2 + NQ^2 + RS^2 = \frac{1}{4}d, \qquad (4)$$

and hence, at least one of the segments MP, NQ, and RS has the length less than or equal to $\frac{\sqrt{d}}{2\sqrt{3}}$. If, for instance, $MP \leq \frac{\sqrt{d}}{2\sqrt{3}}$, then we consider two parallel planes containing the edges AB and CD. Clearly, the distance between these planes is less than or equal to MP and the tetrahedron $ABCD$ is included in the region bounded by the two planes.

Observation. There is an alternative way to obtain the equality (4).

The theorem of the median in the triangle MCD gives

$$4MP^2 = 2\left(MC^2 + MD^2\right) - CD^2.$$

The segments MC and MD are medians in the triangles ABC and ABD, hence

$$4MC^2 = 2\left(BC^2 + AC^2\right) - AB^2$$

and

$$4MD^2 = 2\left(BD^2 + AD^2\right) - AB^2.$$

We obtain

$$4MP^2 = BC^2 + AC^2 + BD^2 + AD^2 - AB^2 - CD^2,$$

and the result follows by adding up the similar two equalities to the above one.

5.4. Note that $(a_n)_{n\geq 0}$ is a strictly increasing sequence and that

$$a_{n+1} = 2a_n - 49,$$

for all n. Suppose that $a_n = pq$, $a_{n+1} = rs$, where p, q, r, s are prime numbers such that $q - p = s - r = d > 0$. Then

$$a_n = p\,(p + d) < r\,(r + d) = a_{n+1},$$

and hence $p < r$. Observe that $p > 2$, since all the terms of the sequence are odd numbers. Because

$$2^n \equiv (-1)^n \pmod 3$$

we deduce that

$$a_n \equiv 1 + (-1)^n \pmod 3,$$

and hence exactly one of any two consecutive terms of the sequence is divisible by 3. It follows that $p = 3$. We obviously have $a_{n+1} < 2a_n$, so

$$r\,(r + d) < 6\,(3 + d).$$

Since $r > 3$, this implies $r < 6$, therefore $r = 5$.
We obtained that $a_n = 3\,(3 + d)$ and $a_{n+1} = 5\,(5 + d)$. Replacing into the recursive relation yields

$$5\,(5 + d) = 6\,(3 + d) - 49,$$

hence $d = 56$.

Finally, we find that the requested terms are $a_7 = 3 \cdot 59$ and $a_8 = 5 \cdot 61$.

Second solution. As before, we observe that a_n is divisible by 3 if and only if n is an odd integer. Also, no term of the sequence is divisible by 2. Therefore, if a_n and a_{n+1} satisfy the requested conditions, then $a_n = 3a$ and $a_{n+1} = (3 + k)\,(a + k)$, where a is a prime number greater than 3 and $3 + k$, $a + k$ are also prime numbers. Since $a_{n+1} = 2a_n - 49$, we obtain

$$(3 + k)\,(a + k) = 6a - 49,$$

or

$$k^2 + k(a+3) + 49 - 3a = 0.$$

If we consider this equality a quadratic equation for k, its discriminant must be a perfect square, since k is an integer number. Therefore, there exists a positive integer t such that

$$(a+3)^2 - 4(49-3a) = a^2 + 18a - 187 = t^2.$$

This is equivalent to

$$(a+9)^2 - t^2 = 268,$$

or

$$(a+9-t)(a+9+t) = 2^2 \cdot 67.$$

Since a is an odd prime, $a+9$ must be even, hence $a+9-t$ and $a+9+t$ are both even. The only possibility is $a+9-t = 2$ and $a+9+t = 2 \cdot 67 = 134$. It follows that $a = 59$ and finally, $n = 7$.

The 6th BMO

The sixth Balkan Mathematical Olympiad for high-school students was held between May 1st and May 7th , 1989, in Split, Yugoslavia. The number of the participating countries remained 5: Bulgaria, Cyprus, Greece, Romania and Yugoslavia.

Problems

6.1. Let n be a positive integer and let $d_1, d_2, \ldots, d_k$ be its divisors, such that

$$1 = d_1 < d_2 < \ldots < d_k = n.$$

Find all values of n for which $k \geq 4$ and $n = d_1^2 + d_2^2 + d_3^2 + d_4^2$.

(Bulgaria)

6.2. Let $\overline{a_n a_{n-1} \ldots a_1 a_0} = a_n 10^n + a_{n-1} 10^{n-1} + \ldots + a_1 10 + a_0$ be the decimal representation of a prime positive integer such that $n > 1$ and $a_n > 1$. Prove that the polynomial

$$P(x) = a_n x^n + a_{n-1} x^{n-1} + \ldots + a_1 x + a_0$$

cannot be written as a product of two nonconstant integer polynomials.

(Yugoslavia)

6.3. Let G be the centroid of a triangle ABC and let d be a line that intersects the sides AB and AC at B_1 and C_1, respectively, such that the points A and G are not separated by d. Prove that

$$[BB_1GC_1] + [CC_1GB_1] \geq \frac{4}{9}[ABC].$$

When does the equality hold?

(Greece)

6.4. The elements of the set $\mathcal{F}$ are some subsets of $\{1, 2, \ldots, n\}$ and satisfy the conditions:

i) if A belongs to $\mathcal{F}$, then A has three elements;

ii) if A and B are distinct elements of $\mathcal{F}$, then A and B have at most one common element.

Let $f(n)$ be the greatest possible number of elements of $\mathcal{F}$. Prove that

$$\frac{n^2-4n}{6} \leq f(n) \leq \frac{n^2-n}{6}.$$

(Romania)

Solutions

6.1. It is not difficult to see that n cannot be an odd integer. Indeed, if n is odd, then all its divisors are odd as well. But the sum of four odd numbers is even, a contradiction.

Therefore n is even and hence $d_2 = 2$. Moreover, since $d_3^2 + d_4^2 = n - 3$, we deduce that one of the numbers d_3 and d_4 is even and the other is odd.

If d_3 is even, then $d_3 = 2d$, for some $d > 1$. But then d is also a divisor of n, which implies that $d = d_2 = 2$ and $d_3 = 4$. This yields

$$n = 21 + d_4^2,$$

so that d_4 is a divisor of 24 greater than 4. The only choice is $d_4 = 7$, but this implies $n = 70$ and, in this case, $d_3 = 5$, a contradiction.

If d_4 is even, then $d_4 = 2d$, where d is again a divisor of n. If $d = d_2 = 2$, then $d_4 = 4$ and consequently $d_3 = 3$, yielding $n = 30$, which is not divisible by 4. It follows that $d = d_3$, and we obtain

$$n = 5 + d_3^2 + 4d_3^2 = 5\left(1 + d_3^2\right).$$

But d_3 divides n and is relatively prime to $1 + d_3^2$, therefore $d_3 = 5$ and

$$n = 1^2 + 2^2 + 5^2 + 10^2 = 130.$$

Since $130 = 2 \cdot 5 \cdot 13$, its least four divisors are indeed $1, 2, 5$, and 10.

6.2. We will use the following result.
Lemma. If α is a complex root of the polynomial

$$P(x) = a_n x^n + \ldots + a_1 x + a_0,$$

then the following inequality holds:

$$|\alpha| \leq 1 + \max_{0 \leq k \leq n} \left|\frac{a_k}{a_n}\right|.$$

Proof. If $|\alpha| \leq 1$, all is clear. Suppose that $|\alpha| > 1$. Because $P(\alpha) = 0$, we have

$$a_n\alpha^n = -\left(a_{n-1}\alpha^{n-1} + \ldots + a_1\alpha + a_0\right),$$

and therefore

$$\begin{aligned}|\alpha|^n &= \left|\frac{a_{n-1}}{a_n}\alpha^{n-1} + \ldots + \frac{a_1}{a_n}\alpha + \frac{a_0}{a_n}\right| \\ &\leq \left|\frac{a_{n-1}}{a_n}\right| |\alpha|^{n-1} + \ldots + \left|\frac{a_1}{a_n}\right| |\alpha| + \left|\frac{a_0}{a_n}\right|.\end{aligned}$$

Let $M = \max\limits_{0 \leq k \leq n} \left|\frac{a_k}{a_n}\right|$. The above inequality then implies

$$|\alpha|^n \leq M\left(|\alpha|^{n-1} + \ldots + |\alpha| + 1\right) = M\frac{|\alpha|^n - 1}{|\alpha| - 1},$$

which easily gives

$$|\alpha|^n\left(|\alpha| - 1 - M\right) \leq -M.$$

Since $M \geq 0$, it follows that $|\alpha| \leq 1 + M$, and the lemma is proved.

Returning to our problem, note that $a_n \geq 2$ and $a_k \leq 9$, for all k, $0 \leq k \leq n-1$. We deduce that $M = \max\limits_{0 \leq k \leq n} \left|\frac{a_k}{a_n}\right| \leq \frac{9}{2}$ and hence, if α is a root of P, then

$$|\alpha| \leq 1 + \frac{9}{2} = \frac{11}{2}.$$

Now, let us assume, by way of contradiction, that $P(x) = Q(x)R(x)$, where Q and R are nonconstant polynomials with integer coefficients. Since

$$P(10) = \overline{a_n a_{n-1} \ldots a_1 a_0} = Q(10)R(10)$$

is a prime number, it follows that one of the numbers $|Q(10)|$ and $|R(10)|$ is equal to 1. Suppose that $|Q(10)| = 1$ and let $x_1, x_2, \ldots, x_k$ be the roots of Q. We then have

$$Q(x) = a(x - x_1)(x - x_2)\ldots(x - x_k),$$

for some integer a and then

$$1 = |a|\,|10 - x_1|\,|10 - x_2|\ldots|10 - x_k|. \tag{1}$$

But $x_1, x_2, \ldots, x_k$ are also roots of P and therefore

$$|10 - x_i| \geq 10 - |x_i| \geq 10 - \frac{11}{2} = \frac{9}{2},$$

for all i, $1 \leq i \leq k$. This shows that equality (1) cannot hold.

Second solution. We prove a stronger result, discarding the condition $a_n > 1$. For this, we use another lemma.

Lemma. If α is a complex root of the polynomial

$$P(x) = a_n x^n + \ldots + a_1 x + a_0,$$

where $\frac{a_{n-1}}{a_n} \geq 0$, then the following inequality holds:

$$\operatorname{Re}\alpha < \frac{1+\sqrt{1+4M}}{2},$$

where $M = \max\limits_{0\leq k\leq n} \left|\frac{a_k}{a_n}\right|$.

Proof. Suppose that α is a root of P such that $\operatorname{Re}\alpha \geq \frac{1+\sqrt{1+4M}}{2}$. Then we also have

$$|\alpha| \geq \operatorname{Re}\alpha \geq \frac{1+\sqrt{1+4M}}{2}.$$

Since $P(\alpha) = 0$, we can write

$$a_n\alpha^n + a_{n-1}\alpha^{n-1} = -\left(a_{n-2}\alpha^{n-2} + \ldots + a_1\alpha + a_0\right),$$

which implies

$$\alpha + \frac{a_{n-1}}{a_n} = -\left(\frac{a_{n-2}}{a_n}\cdot\frac{1}{\alpha} + \frac{a_{n-3}}{a_n}\cdot\frac{1}{\alpha^2} + \ldots + \frac{a_0}{a_n}\cdot\frac{1}{\alpha^{n-1}}\right).$$

We obtain

$$\left|\alpha + \frac{a_{n-1}}{a_n}\right| \leq M\left(\frac{1}{|\alpha|} + \frac{1}{|\alpha|^2} + \ldots + \frac{1}{|\alpha|^{n-1}}\right) < \frac{M}{|\alpha|-1} \leq \frac{2M}{\sqrt{1+4M}-1}.$$

On the other hand,

$$\left|\alpha + \frac{a_{n-1}}{a_n}\right| \geq \operatorname{Re}\left(\alpha + \frac{a_{n-1}}{a_n}\right) = \operatorname{Re}\alpha + \frac{a_{n-1}}{a_n} \geq \operatorname{Re}\alpha \geq \frac{1+\sqrt{1+4M}}{2}.$$

We deduce that

$$\frac{1+\sqrt{1+4M}}{2} < \frac{2M}{\sqrt{1+4M}-1},$$

or

$$4M < 4M,$$

a contradiction. The lemma is proved.

Returning to our problem, let us observe that if α is a complex number such that $\operatorname{Re}\alpha < \frac{19}{2}$, then

$$|10-\alpha| > |9-\alpha|.$$

Indeed, write $\alpha = x+yi$, with $x, y \in \mathbb{R}$. Then the above inequality is equivalent to

$$(10-x)^2 + y^2 > (9-x)^2 + y^2,$$

or

$$x < \frac{19}{2},$$

as stated.

As before, assume, by way of contradiction, that $P(x) = Q(x) R(x)$, where Q and R are nonconstant polynomials with integer coefficients. It follows that one of the numbers $|Q(10)|$ and $|R(10)|$ is equal to 1. Suppose that $|Q(10)| = 1$ and let $x_1, x_2, \ldots, x_k$ be the roots of Q. We then have

$$Q(x) = a(x-x_1)(x-x_2)\ldots(x-x_k),$$

for some integer a and then

$$1 = |a|\,|10-x_1|\,|10-x_2|\ldots|10-x_k|.$$

But $x_1, x_2, \ldots, x_k$ are also roots of P and therefore, by the lemma,

$$\operatorname{Re} x_i < \frac{1+\sqrt{37}}{2} < \frac{19}{2}.$$

Hence

$$|10-x_i| > |9-x_i|,$$

for all i, $1 \le i \le k$. We deduce that

$$1 = |a|\,|10-x_1|\ldots|10-x_k| > |a|\,|9-x_1|\ldots|9-x_k| = |Q(9)|.$$

Clearly $P(9) = Q(9) R(9)$ is a positive integer, therefore $|Q(9)| \ge 1$, which is a contradiction.

6.3. Let D be the midpoint of BC and suppose that AD and B_1C_1 intersect at M (see Figure 6.1).

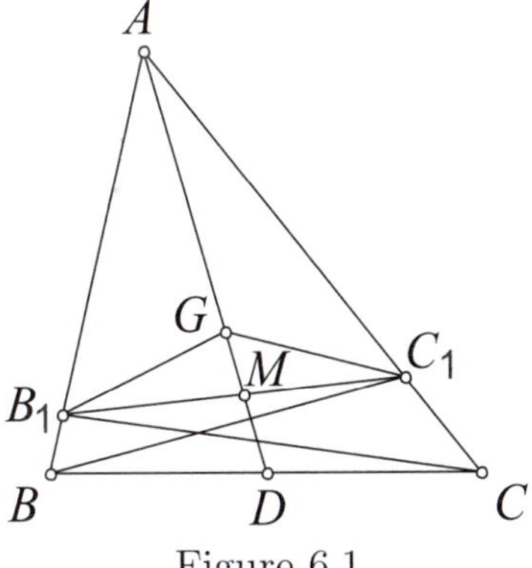

Figure 6.1

Denote

$$\frac{AC_1}{AC} = \alpha, \ \frac{AB_1}{AB} = \beta.$$

We claim that

$$\frac{AM}{AD} = \frac{2\alpha\beta}{\alpha+\beta}.$$

Indeed, let $\frac{AM}{AD} = \gamma$. Then we have

$$\overline{MC_1} = \overline{AC_1} - \overline{AM} = \alpha\overline{AC} - \gamma\overline{AD} = \left(\alpha - \frac{\gamma}{2}\right)\overline{AC} - \frac{\gamma}{2}\overline{AB}.$$

Similarly,

$$\overline{MB_1} = -\frac{\gamma}{2}\overline{AC} + \left(\beta - \frac{\gamma}{2}\right)\overline{AB}.$$

The vectors MB_1 and MC_1 are collinear, therefore

$$\frac{\alpha - \frac{\gamma}{2}}{-\frac{\gamma}{2}} = \frac{-\frac{\gamma}{2}}{\beta - \frac{\gamma}{2}},$$

and this readily gives

$$\gamma = \frac{2\alpha\beta}{\alpha+\beta}.$$

Now, because the points A and G are not separated by the line B_1C_1, it follows that

$$\frac{AM}{AD} \geq \frac{AG}{AD} = \frac{2}{3}$$

and hence

$$\frac{2\alpha\beta}{\alpha+\beta} \geq \frac{2}{3}. \tag{2}$$

Observe that

$$[BB_1GC_1] = [ABC_1] - [AGC_1] - [AGB_1].$$

We have

$$\frac{[ABC_1]}{[ABC]} = \frac{AC_1}{AC} = \alpha,$$

$$\frac{[AGC_1]}{[ABC]} = \frac{[AGC_1]}{2\,[ADC]} = \frac{AG \cdot AC_1}{2AD \cdot AC} = \frac{\alpha}{3}$$

and

$$\frac{[AGB_1]}{[ABC]} = \frac{[AGB_1]}{2\,[ADB]} = \frac{AG \cdot AB_1}{2AD \cdot AB} = \frac{\beta}{3}.$$

It follows that

$$[BB_1GC_1] = \frac{2\alpha - \beta}{3}\,[ABC].$$

Similarly,

$$[CC_1GB_1] = \frac{2\beta - \alpha}{3}[ABC],$$

and thus the requested inequality becomes

$$\frac{\alpha + \beta}{3} \geq \frac{4}{9}.$$

This is easy to prove. Observe that (2) implies

$$\alpha\beta \geq \frac{\alpha + \beta}{3}$$

and then

$$(\alpha + \beta)^2 \geq 4\alpha\beta \geq \frac{4}{3}(\alpha + \beta)$$

which clearly leads to

$$\frac{\alpha + \beta}{3} \geq \frac{4}{9}.$$

The equality holds when $\alpha = \beta$ and $\frac{2\alpha\beta}{\alpha+\beta} = \frac{2}{3}$, that is, when $\alpha = \beta = \frac{2}{3}$. This happens when B_1C_1 is parallel to BC and passes through G.

Second solution. First, observe that

$$[BB_1GC_1] + [CC_1GB_1] = 2[B_1GC_1] + [BB_1C_1] + [CC_1B_1].$$

Projecting the points B, C, and D on the line B_1C_1 to the points B', C', and D', respectively, we have

$$BB' + CC' = 2DD',$$

since DD' is a midline in the trapezoid $BCC'B'$ (see Figure 6.2).

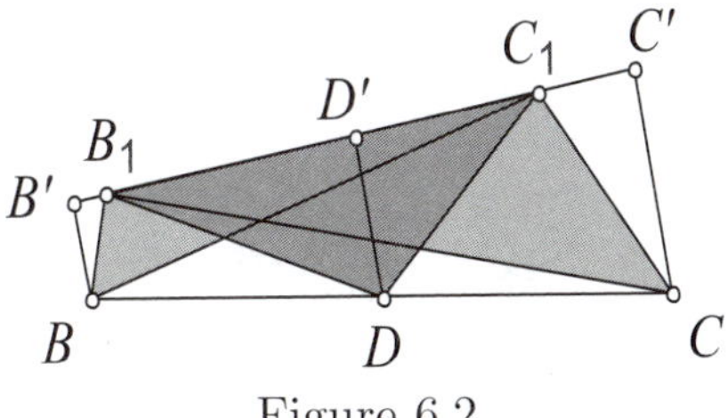

Figure 6.2

Therefore,

$$[BB_1C_1] + [CC_1B_1] = 2[DB_1C_1].$$

So, we have to prove that

$$2[DB_1GC_1] \geq \frac{4}{9}[ABC].$$

Draw through G a parallel to B_1C_1 which intersects the sides AB and AC at the points B_2 and C_2 (see Figure 6.3). Then

$$[DB_1GC_1] \geq [DB_2C_2],$$

since the points B_2 and C_2 are closer to AD than B_1 and C_1, respectively, and therefore $[DB_1G] \geq [DB_2G]$ and $[DC_1G] \geq [DC_2G]$.

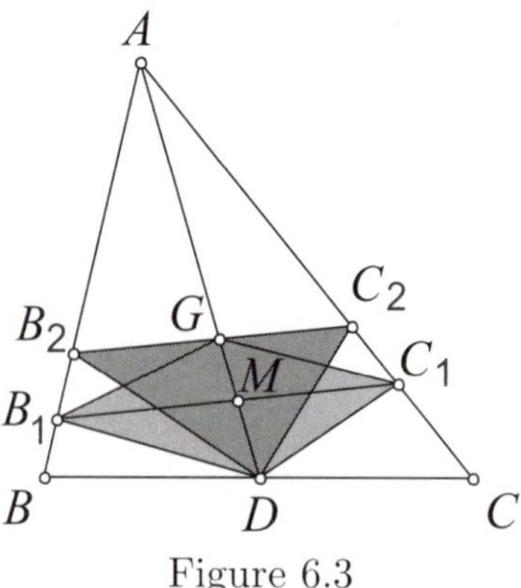

Figure 6.3

Since $AG = 2GD$, we have

$$[AB_2C_2] = 2\,[DB_2C_2].$$

It follows that it suffices to prove that

$$[AB_2C_2] \geq \frac{4}{9}\,[ABC].$$

Draw through G a parallel to BC which intersects the sides AB and AC at B_3 and C_3. It is not difficult to see that

$$[AB_2C_2] \geq [AB_3C_3].$$

Indeed, if B_2 lies between B and B_3 (see Figure 6.4) then

$$[AB_2C_2] - [AB_3C_3] = [TB_2B_3] \geq 0,$$

while if B_2 lies between B_3 and A (see Figure 6.5) then

$$[AB_2C_2] - [AB_3C_3] = [SC_2C_3] \geq 0.$$

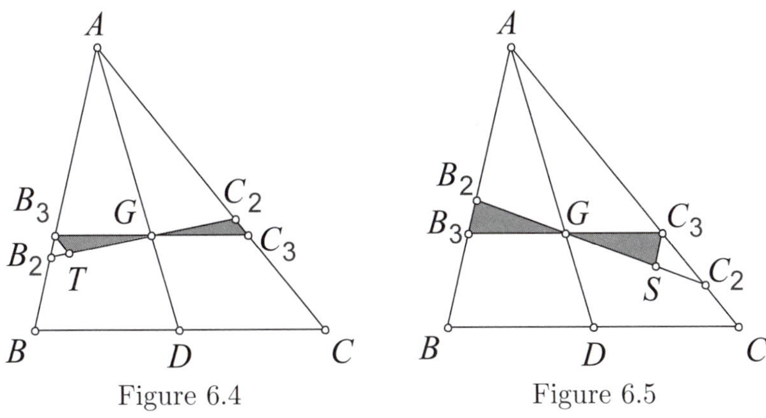

Figure 6.4 Figure 6.5

Since

$$[AB_3C_3] = \frac{4}{9}[ABC],$$

we are done.

6.4. We construct a set $\mathcal{F}_0$ having at least $\frac{n^2-4n}{6}$ elements. This will prove the inequality

$$f(n) \geq \frac{n^2-4n}{6}.$$

Let $\mathcal{F}_1$ be the set of all triples (a, b, c) of distinct elements of $\{1, 2, \ldots, n\}$ such that either

$$a + b + c = n$$

or

$$a + b + c = 2n.$$

Let $\mathcal{F}_0$ be the set of all subsets $\{a, b, c\}$ with $(a, b, c) \in \mathcal{F}_1$. Clearly, $|\mathcal{F}_1| = 6|\mathcal{F}_0|$ and hence it suffices to prove that $|\mathcal{F}_1| \geq n^2-4n$. We estimate now the number of elements of $\mathcal{F}_1$, by counting the number of choices of triples (a, b, c) with the desired property.

The number a can be chosen in n ways. At a first glance, b can be chosen in $n-1$ ways (recall that we must have $b \neq a$) and then c is uniquely determined, namely $c = n-a-b$, if $a+b < n$, or $c = 2n-a-b$, otherwise. However, since $c \neq b$, we must have $b \neq \frac{n-a}{2}$ and $b \neq 2n-2a$ (the second restriction applies only if $a \geq \frac{n}{2}$). Similarly, since $c \neq a$, we impose $b \neq n-\frac{a}{2}$ and $b \neq n-2a$ (the second restriction applies only if $a < \frac{n}{2}$). We conclude that b can be chosen in at least $n-4$ ways, thus $\mathcal{F}_1$ has at least $n(n-4)$ elements.

The final observation is that if $\{a, b, c\}$ and $\{a', b', c'\}$ are distinct elements of $\mathcal{F}_0$, then they have at most one common element. Indeed, if for instance

$a = a'$ and $b = b'$, then $c - c' = \pm n$, which is impossible, because both c and c' belong to the set $\{1, 2, \ldots, n\}$.

For the second inequality, observe that the set $\{1, 2, \ldots, n\}$ contains $\binom{n}{2} = \frac{n(n-1)}{2}$ subsets with two elements. Each element of $\mathcal{F}$ contains exactly three such subsets and no two elements of $\mathcal{F}$ contain the same subset (because any two elements of $\mathcal{F}$ have at most one common element). This proves that

$$f(n) \leq \frac{n(n-1)}{6}.$$

The 7th BMO

The seventh Balkan Mathematical Olympiad for high-school students was held between May 6th and May 11th, 1990, in Sofia, Bulgaria. The number of the participating countries increased to 6: Albania, Bulgaria, Cyprus, Greece, Romania and Yugoslavia.

Problems

7.1. The sequence $(a_n)_{n\geq 1}$ is defined by $a_1 = 1$, $a_2 = 3$ and

$$a_{n+2} = (n+3)\,a_{n+1} - (n+2)\,a_n,$$

for all positive integers n.

Find the values of n for which a_n is divisible by 11.

(Greece)

7.2. The polynomial $P(X)$ is defined by

$$P(X) = \left(X + 2X^2 + \ldots + nX^n\right)^2 = a_0 + a_1 X + \ldots + a_{2n} X^{2n}.$$

Prove that

$$a_{n+1} + a_{n+2} + \ldots + a_{2n} = \frac{n(n+1)\left(5n^2 + 5n + 2\right)}{24}.$$

(Bulgaria)

7.3. Let ABC be an acute angled triangle and let A_1, B_1, and C_1 be the feet of its altitudes. The incircle of the triangle $A_1B_1C_1$ touches its sides at the points A_2, B_2, and C_2. Prove that the Euler lines of triangles ABC and $A_2B_2C_2$ coincide.

(Yugoslavia)

7.4. Find the least number of elements of a finite set A such that there exists a function $f : \{1, 2, 3, ...\} \rightarrow A$ with the property: if i and j are positive integers and $i - j$ is a prime number, then $f(i)$ and $f(j)$ are distinct elements of A.

(Romania)

Solutions

7.1. Let b_n be the reminder of a_n when divided by 11. A short computation shows that $b_1 = 1$, $b_2 = 3$, $b_3 = 9$, $b_4 = 0$, $b_5 = 1$, $b_6 = 7$, $b_7 = 5$, $b_8 = 0$, $b_9 = 1$, $b_{10} = 0$, $b_{11} = 0$. This proves that a_{10} and a_{11} are both divisible by 11 and from the recursive relation we deduce that for all $n \geq 10$, a_n is divisible by 11. Thus, the requested terms are a_4, a_8, and a_n, for all $n \geq 10$.
Observation. We can obtain a formula for the general term a_n. Denoting by $c_n = a_{n+1} - a_n$, we have $c_{n+1} = (n+2)\, c_n$ and since $c_1 = 2$, it follows inductively that $c_n = (n+1)!$. We deduce that

$$a_n = 1! + 2! + \ldots + n!,$$

for all $n \geq 1$.

7.2. We have

$$P(X) = \left(X + 2X^2 + \ldots + nX^n\right)\left(X + 2X^2 + \ldots + nX^n\right),$$

therefore, we obtain

$$\begin{aligned} a_{n+1} &= 1 \cdot n + 2 \cdot (n-1) + 3 \cdot (n-2) + \ldots + (n-1) \cdot 2 + n \cdot 1, \\ a_{n+2} &= 2 \cdot n + 3 \cdot (n-1) + 4 \cdot (n-2) + \ldots + n \cdot 2, \\ &\vdots \\ a_{2n-1} &= (n-1) \cdot n + n \cdot (n-1), \\ a_{2n} &= n^2. \end{aligned}$$

Thus, the requested sum equals

$$\begin{aligned} \sum_{k=1}^{n}\sum_{i=0}^{n-k} (k+i)(n-i) &= \sum_{k=1}^{n}\sum_{i=0}^{n-k} \left(kn + i(n-k) - i^2\right) \\ &= \sum_{k=1}^{n} (n-k+1)\left[kn + \frac{(n-k)^2}{2} - \frac{(n-k)(2n-2k+1)}{6}\right] \\ &= -\frac{1}{6}\sum_{k=1}^{n} k^3 - \frac{n}{2}\sum_{k=1}^{n} k^2 + \frac{3n^2 + 6n + 1}{6}\sum_{k=1}^{n} k + \frac{n^2(n^2-1)}{6} \\ &= \frac{n(n+1)\left(5n^2 + 5n + 2\right)}{24}. \end{aligned}$$

Second solution. Observe that

$$a_0 + a_1 + \ldots + a_{2n} = P(1) = \frac{n^2(n+1)^2}{4}.$$

For $1 \leq k \leq n$ we have

$$a_k = 1 \cdot (k-1) + 2 \cdot (k-2) + \ldots + (k-1) \cdot 1 = \frac{1}{6}\left(k^3 - k\right),$$

so that

$$a_0 + a_1 + \ldots + a_n = \frac{1}{6}\sum_{k=1}^{n} k^3 - \frac{1}{6}\sum_{k=1}^{n} k = \frac{1}{24} n(n+1)\left(n^2 + n - 2\right).$$

Finally, the sum $a_n + a_{n+1} + \ldots + a_{2n}$ equals

$$\frac{n^2(n+1)^2}{4} - \frac{1}{24} n(n+1)\left(n^2+n-2\right) = \frac{n(n+1)\left(5n^2+5n+2\right)}{24}.$$

7.3. Let H be the orthocenter of the triangle ABC. We claim that H is also the incenter of the triangle $A_1B_1C_1$. Observe that $\angle HA_1B = \angle HC_1B = 90°$, therefore the quadrilateral HA_1BC_1 is cyclic. It results that $\angle HBC_1 = \angle HA_1C_1$ (see Figure 7.1).

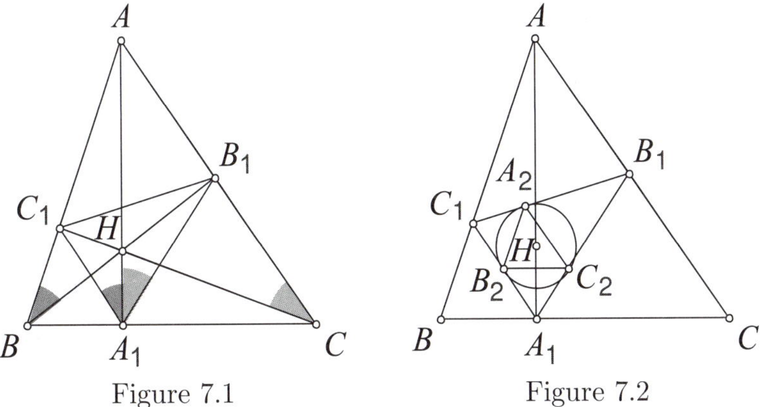

Figure 7.1 Figure 7.2

In a similar way we obtain $\angle HCB_1 = \angle HA_1B_1$. But the quadrilateral BCB_1C_1 is also cyclic (note that $\angle BB_1C = \angle CC_1B$) therefore $\angle C_1BB_1 = \angle C_1CB_1$. We deduce that $\angle HA_1C_1 = \angle HA_1B_1$, that is, HA_1 is the bisector of the angle $\angle B_1A_1C_1$. We conclude that H is the incenter of the triangle $A_1B_1C_1$ and consequently, the circumcenter of the triangle $A_2B_2C_2$ (see Figure 7.2).

It is not difficult to see that $A_1B_2 = A_1C_2$, thus A_1H is perpendicular to B_2C_2. Since A_1H is also perpendicular to BC, it follows that BC and B_2C_2 are parallel. We conclude that the triangles ABC and $A_2B_2C_2$ have parallel sides, so $A_2B_2C_2$ is the image of ABC through an appropriate homothety. It follows that the Euler lines of the two triangles are either parallel or coincide. Since the Euler line of a triangle passes through its circumcenter and its orthocenter, we see that H lies on both Euler lines, hence they coincide.

7.4. Observe that $f(1)$, $f(3)$, $f(6)$, and $f(8)$ must be distinct, since all pairwise differences between the numbers $1, 3, 6$, and 8 are prime numbers. It follows that A has at least 4 elements.

Now, let $A = \{0, 1, 2, 3\}$ and f defined as follows: $f(n)$ is the remainder of n when divided by 4. If $f(i) = f(j)$, then $i - j$ is divisible by 4. Therefore, if $i - j$ is a prime number, then $f(i) \neq f(j)$. We conclude that the requested number of elements is 4.

The 8th BMO

The eighth Balkan Mathematical Olympiad for high-school students was held between May 13th and May 16th , 1991, in Constanta, Romania. The number of the participating countries remained 6: Albania, Bulgaria, Cyprus, Greece, Romania and Yugoslavia.

Problems

8.1. Let ABC be an acute triangle inscribed in a circle centered at O. Let M be a point on the small arc AB of the triangle's circumcircle. The perpendicular dropped from M on the ray OA intersects the sides AB and AC at the points K and L, respectively. Similarly, the perpendicular dropped from M on the ray OB intersects the sides AB and BC at N and P, respectively. Assume that $KL = MN$. Find the size of the angle $\angle MLP$ in terms of the angles of the triangle ABC.

(Greece)

8.2. Show that there are infinitely many noncongruent triangles which satisfy the following conditions:

i) the side lengths are relatively prime integers;

ii) the area is an integer number;

iii) the altitudes' lengths are not integer numbers.

(Yugoslavia)

8.3. A regular hexagon of area H is inscribed in a convex polygon of area P. Show that $P \leq \frac{3}{2}H$. When does the equality hold?

(Bulgaria)

8.4. Prove that there is no bijective function $f : \{1, 2, 3, \ldots\} \to \{1, 2, 3, \ldots\}$ such that

$$f(mn) = f(m) + f(n) + 3f(m)f(n),$$

for all positive integers m and n.

(Romania)

Solutions

8.1. Since $\angle AOB = 2\angle ACB$, it follows that

$$\angle OAB = \angle OBA = 90^\circ - \angle C.$$

The line KL is perpendicular to AO, so

$$\angle AKL = 90^\circ - \angle OAB = \angle C.$$

Similarly, $\angle BNP = \angle C$. Note that $\angle AKL = \angle MKN$ and $\angle BNP = \angle MNK$. We derive that

$$\angle MKN = \angle MNK = \angle C$$

and hence the triangle MNK is isosceles (see Figure 8.1).

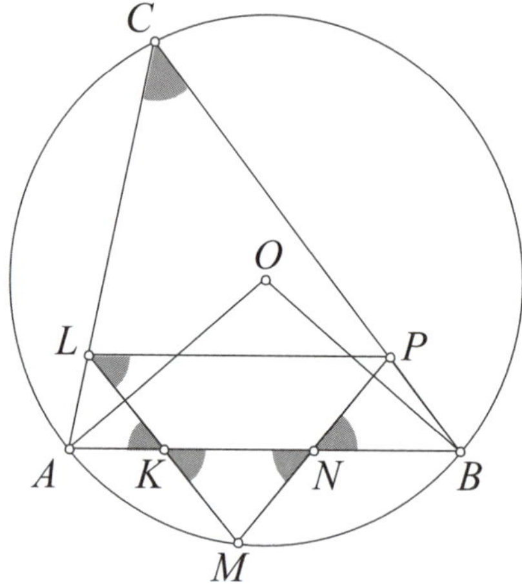

Figure 8.1

This implies that

$$MK = MN. \tag{1}$$

The triangles AKL and ACB share a common angle and $\angle AKL = \angle ACB$; therefore they are similar. This yields

$$\frac{AK}{KL} = \frac{AC}{BC}. \tag{2}$$

In the same way, triangles PBN and ABC are similar, hence

$$\frac{NP}{NB} = \frac{AC}{BC}. \tag{3}$$

Let $\angle MAB = \alpha$ and $\angle MBA = \beta$. Then $\alpha + \beta = \angle C$ and we have

$$\angle AMK = 180^\circ - \alpha - (180^\circ - \angle C) = \angle C - \alpha = \beta.$$

Similarly, we obtain

$$\angle BMN = 180^\circ - \beta - (180^\circ - \angle C) = \angle C - \beta = \alpha,$$

therefore the triangles AKM and MNB are also similar. It results

$$\frac{AK}{KM} = \frac{MN}{NB}. \tag{4}$$

Using the equalities (1)-(4) we successively have

$$\frac{MN}{NP} = \frac{MK \cdot BC}{AC \cdot NB} = \frac{BC \cdot AK}{AC \cdot MN} = \frac{LK}{MN}.$$

From the hypothesis we have $KL = MN$, hence $MN = NP$. Also, (1) yields $KL = MK$, so that KN is a midline in the triangle MLP. Consequently, KN and LP are parallel and

$$\angle MLP = \angle MKN = \angle ACB.$$

8.2. Let S denote the area and p denote the semiperimeter of a triangle with side lengths a, b, c. By Heron's formula, we have

$$S = \sqrt{p(p-a)(p-b)(p-c)}.$$

Also, note that

$$(p-a) + (p-b) + (p-c) = p.$$

We start with the identity

$$n^4 + 4n^2 + 4 = \left(n^2 + 2\right)^2,$$

and we define the triangle T_n by setting $p - a = n^4$, $p - b = 4n^2$, $p - c = 4$. Thus, $p = \left(n^2 + 2\right)^2$ and $S = 4n^3\left(n^2 + 2\right)$. Condition ii) is fulfilled.

The side lengths of T_n are $a = 4n^2 + 4 = 4\left(n^2 + 1\right)$, $b = n^4 + 4$, and $c = n^4 + 4n^2 = n^2\left(n^2 + 4\right)$. It is easy to check that $a + b > c$, $b + c > a$, and $a + c > b$, therefore there exists a triangle with side lengths a, b, c.

We claim that if n is an odd integer, then $\gcd(a, c) = 1$. Indeed, let d be a prime number dividing both a and c. Since c is odd, $d > 2$. Then d divides both $n^2 + 1$ and $n^2 + 4$, hence $d = 3$. It follows that $n^2 \equiv -1 \pmod 3$ and this is a contradiction. It follows that for odd integers n, $\gcd(a, b, c) = 1$ and the triangles T_n satisfy condition i)

The altitudes of T_n are given by the formulae

$$h_a = \frac{2S}{a} = \frac{2n^3\left(n^2+2\right)}{n^2+1},$$
$$h_b = \frac{2S}{b} = \frac{8n^3\left(n^2+2\right)}{n^4+4},$$
$$h_c = \frac{2S}{c} = \frac{8n\left(n^2+2\right)}{n^2+4}.$$

If n is odd and $n > 1$, the above rational numbers are represented by fractions which can be reduced only by dividing by 2. Thus, condition iii) is also fulfilled.

Second solution. Other families of triangles $(T_n)_n$ can be considered as solutions to the problem. For example:

$$a = (n+1)^4\,,\ b = n^2\left[(n+1)^2+1\right],\ c = n^2+(n+1)^2\,,$$

where $n \geq 5$ is an odd integer.

Another example can be given by taking

$$a = 4n^2+1,\ b = 4n^4+1,\ c = 4n^2\left(n^2+1\right),$$

which can be obtained as in the first solution considering the identity

$$\left(2n^2+1\right)^2 = 4n^4+4n^2+1.$$

For the altitudes we obtain the formulae

$$h_a = \frac{8n^3\left(n^2+1\right)}{4n^2+1},\ h_b = \frac{8n^3\left(2n^2+1\right)}{4n^4+1},\ h_c = \frac{2n\left(2n^2+1\right)}{n^2+1}.$$

8.3. Let $A_1A_2A_3A_4A_5A_6$ be the regular hexagon and let O be its center. First, notice that we can include the convex polygon in a greater hexagon $C_1C_2\ldots C_6$. Indeed, if X is a point on the border of a convex polygon, there exists at least one line l_X passing through X such that the polygon is included in one of the halfplanes determined by it. Draw such lines $l_1, l_2, \ldots, l_6$ passing through the points $A_1, \ldots, A_6$, determining thus the hexagon $C_1C_2\ldots C_6$ (see Figure 8.2). It suffices to prove that the area of this new polygon is no greater than $\frac{3}{2}H$.

Let B_1 be the point of intersection between the lines A_1A_6 and A_2A_3 and let B_2 be the point of intersection between the lines A_1A_2 and A_3A_4. In the same way we define the points B_3, B_4, B_5, and B_6.

We can see that the hexagon $A_1A_2\ldots A_6$ appears as the intersection of the equilateral triangles $B_1B_3B_5$ and $B_2B_4B_6$ (see Figure 8.3).

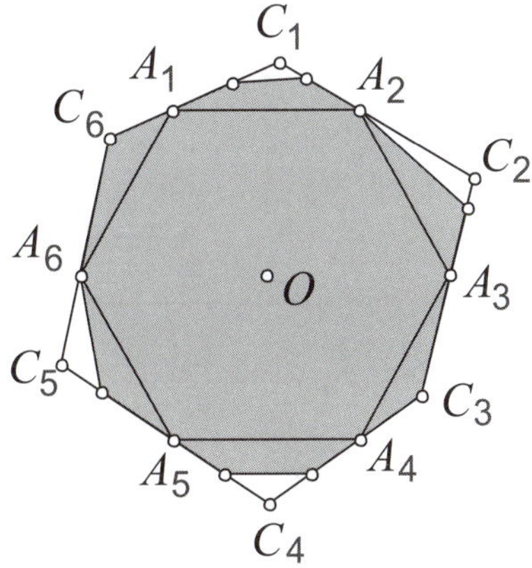

Figure 8.2

We denote by s the area of the small equilateral triangle OA_1A_2. Then $[A_1B_1A_2] = s$ and $H = 6s$.

Let $p_1, p_2, \ldots, p_6$ be the parts of the convex hexagon $C_1C_2 \ldots C_6$ which lie outside the regular hexagon and are bounded by the sides A_1A_2, $A_2A_3, \ldots, A_6A_1$, respectively. It is sufficient to prove that

$$p_1 + p_2 + \ldots + p_6 \leq \frac{1}{2}H = 3s.$$

For this, we will prove that $p_1 + p_2 \leq s$. Adding up the similar inequalities will give the desired result.

Observe that since the polygon is convex, C_1 and C_2 must lie in the interior or on the sides of the triangles $A_1B_1A_2$ and $A_2B_2A_3$.

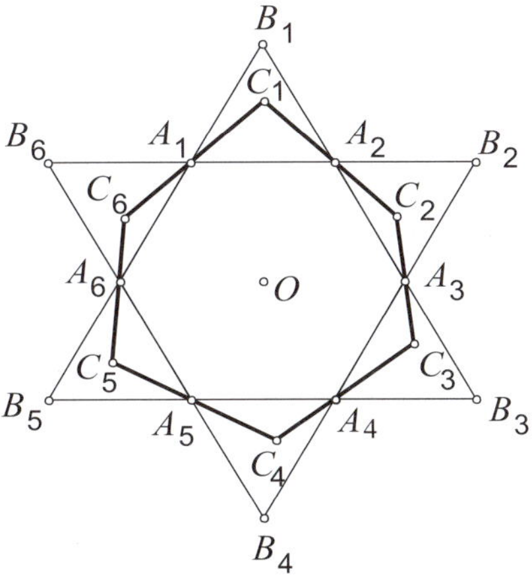

Figure 8.3

The central symmetry with center A_2 sends A_1 to B_2 and C_1 to a point D lying on the ray A_2C_2 (see Figure 8.4). Therefore the triangles $A_2C_2A_3$ and A_2DB_2 do not overlap and the sum of their areas do not exceed the area of triangle $A_2B_2A_3$. Since the triangle A_2DB_2 is the image of $A_2C_1A_1$ through the symmetry, we deduce that $p_1 + p_2 \leq s$, as desired.

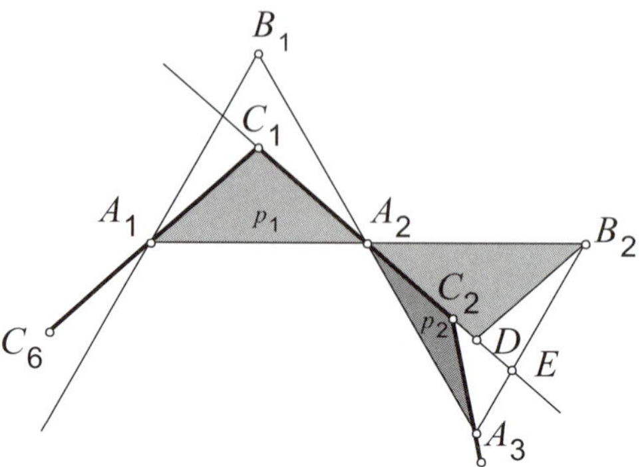

Figure 8.4

If the equality occurs, then we must have $[B_2DE] = 0$ (the point E is the intersection between C_1C_2 and A_3B_2). It is not difficult to see that this happens either if C_1 lies on A_1B_1 and C_2 on A_2A_3 or if C_1 lies on A_1A_2 and C_2 on B_2A_3. Extending this to the other vertices, we deduce that the equality occurs if the convex polygon coincides with either $B_1B_3B_5$ or $B_2B_4B_6$.

Second solution. We use the following result.
Lemma. The length of a triangle's side equals l and its adjacent angles are α and β. If the area of the triangle is S, then

$$S \leq \frac{l^2}{8}(\tan\alpha + \tan\beta).$$

Proof. Denote by l_α and l_β the lengths of the other two sides of the triangle. The sine law gives

$$\frac{l_\alpha}{\sin\alpha} = \frac{l_\beta}{\sin\beta} = \frac{l}{\sin(180^\circ - \alpha - \beta)} = \frac{l}{\sin(\alpha+\beta)}.$$

Using this, we have

$$S = \frac{1}{2}l_\alpha l_\beta \sin(\alpha+\beta) = \frac{1}{2}l^2\frac{\sin\alpha\sin\beta}{\sin(\alpha+\beta)}.$$

But

$$\sin(\alpha+\beta) = \sin\alpha\cos\beta + \sin\beta\cos\alpha,$$

hence

$$S = \frac{l^2}{2} \cdot \frac{1}{\cot \alpha + \cot \beta}.$$

We have to prove that

$$\frac{1}{\cot \alpha + \cot \beta} \leq \frac{1}{4} (\tan \alpha + \tan \beta),$$

but this is an easy case of the $HM - AM$ inequality. Observe that the equality holds if and only if $\alpha = \beta$.
Returning to the problem, in each triangle $A_iC_iA_{i+1}$ denote$\angle A_{i+1}C_iA_i = \alpha_i$ (here and in the rest of the solution, $A_7 = A_1$, $\alpha_7 = \alpha_1$, etc.) With no loss of generality, we can assume that the side length of the hexagon $A_1A_2 \ldots A_6$ equals 1, and therefore $s = \frac{\sqrt{3}}{4}$. Using the lemma, we have

$$p_i = [A_iC_iA_{i+1}] \leq \frac{1}{8} (\tan \alpha_i + \tan (60^\circ - \alpha_{i+1})),$$

for all i.

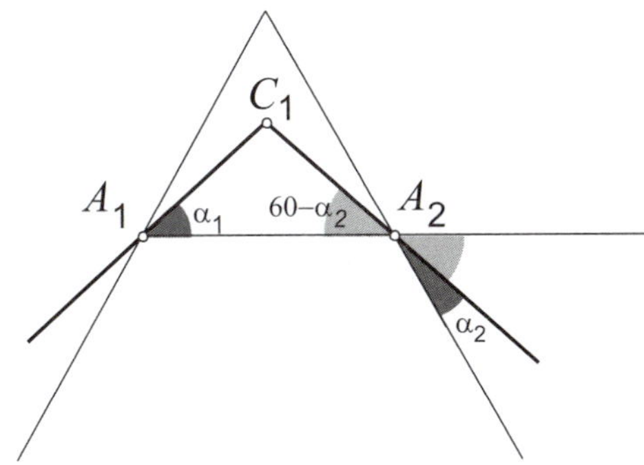

Figure 8.5

Therefore

$$\sum_{i=1}^{6} p_i \leq \frac{1}{8} \sum_{i=1}^{6} (\tan \alpha_i + \tan (60^\circ - \alpha_{i+1})),$$

and the equality holds if $\alpha_i = 60^\circ - \alpha_{i+1}$, for all i. Rearranging terms, we obtain

$$\sum_{i=1}^{6} p_i \leq \frac{1}{8} \sum_{i=1}^{6} (\tan \alpha_i + \tan (60^\circ - \alpha_i)) \tag{1}$$

Observe that if x satisfies $0 \leq x \leq 60^\circ$, then

$$\tan x + \tan (60^\circ - x) \leq \tan 60^\circ. \tag{2}$$

Indeed, we have

$$\tan(60^\circ - x) = \frac{\tan 60^\circ - \tan x}{1 + \tan 60^\circ \tan x} \le \tan 60^\circ - \tan x.$$

Moreover, the equality holds if x equals either 0 or 60°.

Using (1) and (2) we deduce

$$\sum_{i=1}^{6} p_i \le \frac{1}{8} \cdot 6 \tan 60^\circ = \frac{3\sqrt{3}}{4} = 3s,$$

and, as we saw in the previous solution, this implies $P \le \frac{3}{2}H$.

If $P = \frac{3}{2}H$, then the equality must hold in all inequalities. This means that $\alpha_i = 60^\circ - \alpha_{i+1}$, for all i, and for each i, α_i equals either 0 or 60°. We obtain that either $\alpha_1 = \alpha_3 = \alpha_5 = 60^\circ$ and $\alpha_2 = \alpha_4 = \alpha_6 = 0$, or $\alpha_1 = \alpha_3 = \alpha_5 = 0$ and $\alpha_2 = \alpha_4 = \alpha_6 = 60^\circ$, that is, the convex polygon equals either $B_1B_3B_5$ or $B_2B_4B_6$.

8.4. Assume that such a function exists. Taking $m = n = 1$ in the given relation yields

$$f(1) = 2f(1) + 3f^2(1),$$

and hence $f(1) = 0$. Therefore, $f(n) \ge 1$ for all $n \ge 2$. Observe that if a is a composite number, then $a = mn$, for some $m, n \ge 2$. But then

$$f(a) = f(mn) = f(m) + f(n) + 3f(m)f(n) \ge 1 + 1 + 3 = 5.$$

Since f is an onto function, for every number $k \ge 1$, there exists n_k such that $f(n_k) = k$. We have $f(n_1) = 1 < 5$ and $f(n_3) = 3 < 5$, hence n_1 and n_3 are prime numbers. Moreover,

$$f\left(n_3^2\right) = 3 + 3 + 3 \cdot 3 \cdot 3 = 33,$$
$$f(n_1 n_8) = 1 + 8 + 3 \cdot 1 \cdot 8 = 33.$$

But f is one-to-one, therefore $n_3^2 = n_1 n_8$, which is a contradiction, since n_1 and n_3 are distinct primes.

We conclude that no such function exists.

Observation. The reader trained in higher algebra will notice that, essentially, the problem states that the multiplicative monoids $\mathbb{N}^*$ and $3\mathbb{N} + 1$ are not isomorphic. Indeed, let $\varphi : \mathbb{N} \to 3\mathbb{N} + 1$, $\varphi(n) = 3n + 1$ and observe that the composite function $g : \mathbb{N}^* \to 3\mathbb{N} + 1$, $g = \varphi \circ f$ satisfies

$$g(mn) = 3f(mn) + 1 = 3\left(f(m) + f(n) + 3f(m)f(n)\right) + 1$$
$$= (3f(m) + 1)(3f(n) + 1) = g(m)g(n).$$

Moreover, g is a bijective function. We conclude that if such a function f exists, then g is an isomorphism.

It is known that $\mathbb{N}^*$ is a monoid with unique factorization. On the other hand, $3\mathbb{N}+1$ does not have this property. Indeed, we have, for instance,

$$10 \cdot 10 = 4 \cdot 25,$$

and all factors $10, 4$, and 25 are indecomposable elements of the monoid $3\mathbb{N}+1$.

The 9th BMO

The ninth Balkan Mathematical Olympiad for high-school students was held between May 4th and May 9th, 1992, in Athens, Greece. The number of the participating countries remained 6: Albania, Bulgaria, Cyprus, Greece, Romania and Yugoslavia.

Problems

9.1. Let m and n be positive integers and

$$A(m,n) = m^{3^{4n}+6} - m^{3^{4n}+4} - m^5 + m^3.$$

Find all integers n with the property that $A(m,n)$ is divisible by 1992 for each integer m.

(Bulgaria)

9.2. Show that for every positive integer n the following inequality holds:

$$\left(2n^2 + 3n + 1\right)^n \geq 6^n (n!)^2.$$

(Cyprus)

9.3. Let ABC be a triangle and let D, E, F be points in the interior of the sides BC, CA, and AB, respectively, such that the quadrilateral $AFDE$ is cyclic. Prove that

$$4\frac{[DEF]}{[ABC]} \leq \left(\frac{EF}{AD}\right)^2.$$

9.4. For each positive integer $n \geq 3$, find the least positive integer $f(n)$ with the following property: for any subset $A \subset \{1, 2, \ldots, n\}$ with $f(n)$ elements, there exist $x, y, z \in A$ which are pairwise prime.

(Romania)

Solutions

9.1. We have the factorizations

$$1992 = 2^3 \cdot 3 \cdot 83$$

and

$$A(m,n) = (m-1)\, m^3 \,(m+1)\left(m^{3^{4n}+1} - 1\right).$$

Observe that $(m-1)\, m\,(m+1)$ is divisible by 3 for all integers m. Also, if m is even, 2^3 divides m^3 and if m is odd, then 2^3 divides $m^2 - 1$. Therefore, $A(m,n)$ is divisible by $2^3 \cdot 3$ for all m.

Consequently, we have to determine all integers n such that

$$m^{3^{4n}+1} - 1 \equiv 0 \pmod{83}$$

for all positive integers m such that $m \not\equiv 0, 1, -1 \pmod{83}$. As a particular case, the above congruence must hold for $m = 2$. This fails if n is even. Indeed, since $\gcd(2, 83) = 1$, Fermat's little theorem yields

$$2^{82} \equiv 1 \pmod{83},$$

hence

$$2^{82k} \equiv 1 \pmod{83},$$

for all positive integers k. If n is even, then

$$3^{4n} = 81^n = (82-1)^n \equiv 1 \pmod{82},$$

so $3^{4n} = 82k + 1$, for some integer k. We obtain

$$2^{3^{4n}+1} - 1 = 2^{82k+2} - 1 \equiv 3 \pmod{83}.$$

Finally, assume that n is odd. Then, if m is not divisible by 83, we have

$$m^{82} \equiv 1 \pmod{83},$$

so that

$$m^{82k} \equiv 1 \pmod{83},$$

for all k. Also,

$$3^{4n} + 1 = 81^n + 1 = (82-1)^n + 1 \equiv 0 \pmod{82},$$

and hence $3^{4n} + 1 = 82k$, for some integer k. It follows that

$$m^{3^{4n}+1} - 1 \equiv 0 \pmod{83}.$$

We conclude that $A(m,n)$ is divisible by 1992 for all m if and only if n is an odd integer.

9.2. The $AM - GM$ inequality yields

$$\frac{1^2+2^2+\ldots+n^2}{n} \geq \left(1^2 2^2 \ldots n^2\right)^{\frac{1}{n}}.$$

But

$$\frac{1^2+2^2+\ldots+n^2}{n} = \frac{n(n+1)(2n+1)}{6n} = \frac{2n^2+3n+1}{6},$$

and, consequently,

$$\left(2n^2+3n+1\right)^n \geq 6^n (n!)^2,$$

as desired.

Observe that the equality cannot hold unless $n=1$.

9.3. Because $AFDE$ is a cyclic quadrilateral, we have $\angle BAD = \angle FED$ and $\angle DAC = \angle DFE$ (see Figure 9.1). Therefore

$$\frac{[ABD]}{[DEF]} = \frac{AB \cdot AD \cdot \sin \angle BAD}{DE \cdot DF \cdot \sin \angle FED} = \frac{AB \cdot AD}{DE \cdot DF}.$$

Similarly,

$$\frac{[ADC]}{[DEF]} = \frac{AC \cdot AD}{DF \cdot EF}.$$

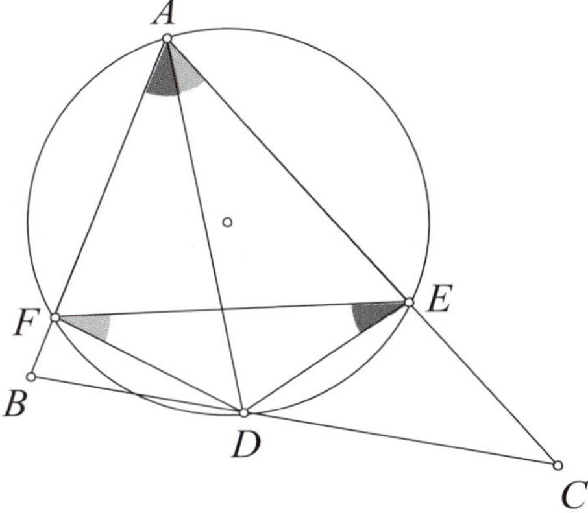

Figure 9.1

Now, we have

$$\frac{[ABC]}{[DEF]} = \frac{[ABD]}{[DEF]} + \frac{[ADC]}{[DEF]} = \frac{AB \cdot AD}{DE \cdot EF} + \frac{AC \cdot AD}{DF \cdot EF} = \frac{AD}{EF}\left(\frac{AB}{DE} + \frac{AC}{DF}\right),$$

and the $AM - GM$ inequality gives

$$\frac{[ABC]}{[DEF]} \geq \frac{AD}{EF} \cdot 2\sqrt{\frac{AB \cdot AC}{DE \cdot DF}}.$$

But

$$\frac{AB \cdot AC}{DE \cdot DF} = \frac{AB \cdot AC \cdot \sin \angle BAC}{DE \cdot DF \cdot \sin \angle FDE} = \frac{[ABC]}{[DEF]}.$$

It follows that

$$\frac{[ABC]}{[DEF]} \geq \frac{AD}{EF} \cdot 2\sqrt{\frac{[ABC]}{[DEF]}},$$

which readily simplifies to the requested inequality.

Second solution. Consider the points B' and C' on the rays $(AB$ and $(AC$ such that D is the midpoint of the line segment $B'C'$ (see Figure 9.2). If, for instance, $BD \leq CD$, then $[BB'D] \leq [CC'D]$ (because $[BB'D] = [C''C'D]$) and hence $[ABC] \geq [AB'C']$. It is therefore sufficient to prove the assertion of the problem for the triangle $AB'C'$ instead of ABC.

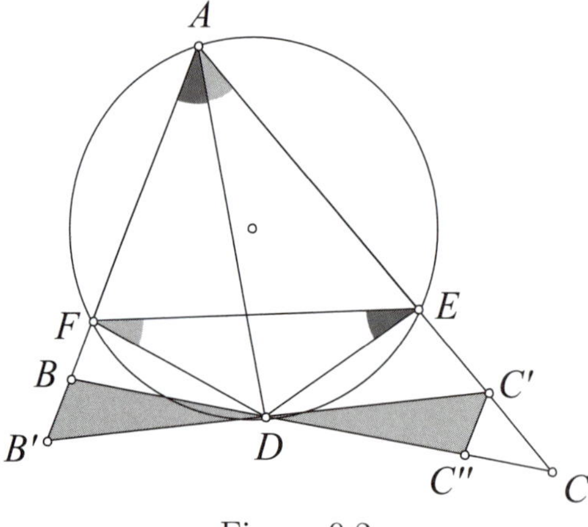

Figure 9.2

We claim that in this case the equality is fulfilled, that is,

$$4\frac{[DEF]}{[AB'C']} = \left(\frac{EF}{AD}\right)^2.$$

Indeed, let M and N be the midpoints of the sides AB' and AC', respectively (see Figure 9.3). Then

$$\angle DAF = \angle DEF$$

and

$$\angle DFE = \angle DAE = ADM.$$

It follows that the triangle DEF and MAD are similar, so we obtain

$$\frac{[DEF]}{[MAD]} = \left(\frac{EF}{AD}\right)^2.$$

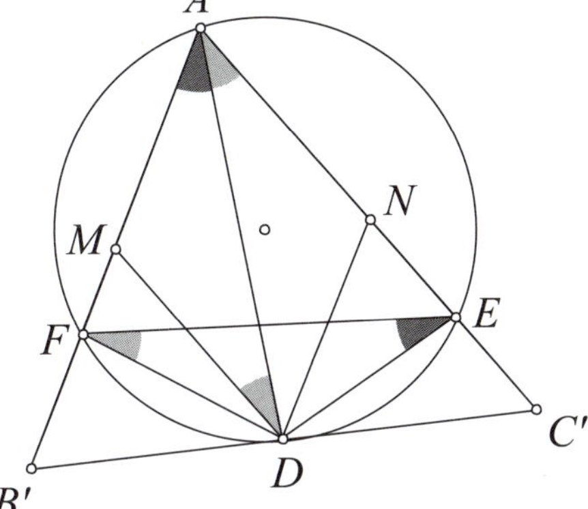

Figure 9.3

Finally, observe that

$$[AB'C'] = 2\,[AB'D] = 4\,[MAD],$$

thus

$$4\frac{[DEF]}{[AB'C']} = \frac{[DEF]}{[MAD]} = \left(\frac{EF}{AD}\right)^2,$$

as desired.

Observation. The construction of the points B' and C' can be done as follows: draw through D parallels to the sides AB and AC, and let M and N be their intersections with the respective sides (see Figure 9.4).

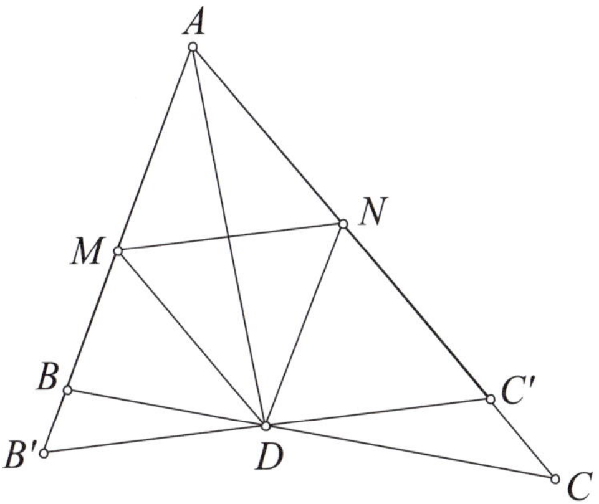

Figure 9.4

Since $AMDN$ is a parallelogram, AD passes through the midpoint of MN. Now draw a line through D, parallel to MN. Its intersections with the lines AB and AC are the requested points B' and C'.

9.4. We claim that

$$f(n) = \left\lfloor \frac{n}{2} \right\rfloor + \left\lfloor \frac{n}{3} \right\rfloor - \left\lfloor \frac{n}{6} \right\rfloor + 1,$$

for all $n \geq 3$.

Consider the set

$$X = \{x \mid 1 \leq x \leq n \text{ and } \gcd(x, 6) \neq 1\}.$$

The inclusion-exclusion principle shows that X has $\lfloor \frac{n}{2} \rfloor + \lfloor \frac{n}{3} \rfloor - \lfloor \frac{n}{6} \rfloor$ elements. Clearly, in any 3-element subset of X we can find two elements which are both divisible by either 2 or 3. This implies that

$$f(n) \geq \left\lfloor \frac{n}{2} \right\rfloor + \left\lfloor \frac{n}{3} \right\rfloor - \left\lfloor \frac{n}{6} \right\rfloor + 1.$$

In order to prove the reversed inequality we need a preliminary result.
Lemma. Let a be a positive integer. Then any 5-element subset A,

$$A \subset \{a, a+1, a+2, a+3, a+4, a+5\}$$

contains three numbers which are pairwise relatively prime.
Proof. Let x be the odd number among $a, a+1$. Then

$$\{x, x+2, x+4\} \subset \{a, a+1.a+2, a+3, a+4, a+5\}$$

and $\gcd(x, x+2) = \gcd(x, x+4) = \gcd(x+2, x+4) = 1$. If y belongs to $\{x+1, x+3\}$ and 3 does not divide y, then $x, x+2, x+4, y$ are also pairwise relatively prime. It follows that for any choice of a 5-element subset A, we can choose at least three among $x, x+2, x+4, y$ which are pairwise relatively prime. The lemma is proved.

We prove now that

$$f(n) \leq \left\lfloor \frac{n}{2} \right\rfloor + \left\lfloor \frac{n}{3} \right\rfloor - \left\lfloor \frac{n}{6} \right\rfloor + 1,$$

by induction on n.

The inequality is true for $3 \leq n \leq 8$. Indeed, for $n = 3$ and $n = 4$ we have $1, 2, 3$ pairwise relatively prime and this proves $f(3) \leq 3$, $f(4) \leq 4$.

For $n = 5$ we need to show that $f(5) \leq 4$. If number 1 was not already chosen, then $\{3, 4, 5\} \subset \{2, 3, 4, 5\}$ will do. If 1 has been chosen, we need to find other two elements among $\{2,3,4,5\}$. Any of the choices $\{2,3\}$ or $\{4,5\}$ works, since $1, x, x+1$ are pairwise relatively prime.

For $n = 6$, we have to prove that $f(6) \leq \lfloor \frac{6}{2} \rfloor + \lfloor \frac{6}{3} \rfloor - \lfloor \frac{6}{6} \rfloor + 1 = 5$. This follows from the lemma.

For $n = 7$, we have to prove that $f(7) \leq 5$. Indeed, if five elements are chosen among $\{2,3,4,5,6,7\}$, then we apply the lemma. In other cases, we choose 1 and other four elements among $\{2,3,4,5,6,7\}$; this gives 1 and two consecutive numbers, which are pairwise relatively prime.

For $n = 8$, we have to prove that $f(8) \leq \lfloor \frac{8}{2} \rfloor + \lfloor \frac{8}{3} \rfloor - \lfloor \frac{8}{6} \rfloor + 1 = 6$. If we choose five elements among $\{3,4,5,6,7,8\}$, the result is given by the lemma. Otherwise, we choose $1, 2$, and other 4 numbers from $\{3,4,5,6,7,8\}$. Then we have chosen at least two consecutive numbers which, together with 1, are pairwise relatively prime.

To end the proof, denote by

$$g(n) = \left\lfloor \frac{n}{2} \right\rfloor + \left\lfloor \frac{n}{3} \right\rfloor - \left\lfloor \frac{n}{6} \right\rfloor + 1,$$

for each $n \geq 3$. It is clear that

$$g(n+6) = g(n) + 4.$$

Assume that $n \geq 9$ and $f(n-6) \leq g(n-6)$. Consider a subset A of the set $\{1, 2, \ldots, n\}$ which contains $g(n)$ elements. If

$$|A \cap \{n-5, n-4, n-3, n-2, n-1, n\}| \geq 5,$$

then, by the lemma, A contains three elements pairwise relatively prime. Otherwise, A contains at least $g(n) - 4 = g(n-6)$ elements from the set $\{1, 2, \ldots, n-6\}$ and, by the induction hypothesis, it contains a 3-element subset with the required property.

The 10th BMO

The tenth Balkan Mathematical Olympiad for high-school students was held between May 3rd and May 8th, 1993, in Nicosia, Cyprus. The number of the participating countries remained 6: Albania, Bulgaria, Cyprus, Greece, Romania and Yugoslavia.

Problems

10.1. The real numbers $a, b, c, d, e,$ and f satisfy the conditions

$$a + b + c + d + e + f = 10$$

and

$$(a-1)^2 + (b-1)^2 + (c-1)^2 + (d-1)^2 + (e-1)^2 + (f-1)^2 = 6.$$

Determine the greatest possible value of f.

(Cyprus)

10.2. A positive integer is called *monotonic* if the digits of its decimal representation

$$a_N 10^N + a_{N-1}10^{N-1} + \ldots + a_1 10 + a_0$$

satisfy $a_N \leq a_{N-1} \leq \ldots \leq a_1 \leq a_0$. Find the number of all monotonic positive integers having at most 1993 digits.

(Bulgaria)

10.3. The circles $\mathcal{C}_1$ and $\mathcal{C}_2$, centered at O_1 and O_2, are externally tangent at T. A third circle $\mathcal{C}$, centered at O, is tangent to $\mathcal{C}_1$ and $\mathcal{C}_2$ at A and B, respectively, such that O_1 and O_2 lie in its interior. The line tangent to $\mathcal{C}_1$ and $\mathcal{C}_2$ at T intersects $\mathcal{C}$ at K and L. Let D be the midpoint of KL. Prove that $\angle O_1OO_2 = \angle ADB$.

(Greece)

10.4. Let p be a prime number and let m be a positive integer. Prove that the equation

$$\frac{x^p + y^p}{2} = \left(\frac{x+y}{2}\right)^m$$

has a solution $(x, y) \neq (1, 1)$ in positive integers if and only if $m = p$.

(Romania)

Solutions

10.1. Observe that

$$6 = \sum (a-1)^2 = \sum a^2 - 2\sum a + 6 = \sum a^2 - 14,$$

hence

$$\sum a^2 = 20.$$

We have

$$a + b + c + d + e = 10 - f$$

and

$$a^2 + b^2 + c^2 + d^2 + e^2 = 20 - f^2.$$

The Cauchy-Schwartz inequality gives

$$5\left(a^2 + b^2 + c^2 + d^2 + e^2\right) \geq (a + b + c + d + e)^2,$$

therefore we obtain

$$5\left(20 - f^2\right) \geq (10 - f)^2,$$

or, equivalently,

$$3f^2 - 10f \leq 0.$$

The latter yields

$$0 \leq f \leq \frac{10}{3},$$

and we see that f can reach the upper bound when $a = b = c = d = e = \frac{4}{3}$.

10.2. We have to count how many finite sequences (x_n) with 1993 terms $x_k \in \{0, 1, \ldots, 9\}$ satisfy the condition

$$x_1 \leq x_2 \leq \ldots \leq x_{1993}.$$

Let X be the set of such sequences. Let Y be the set of sequences (y_n) with 1993 terms $y_k \in \{0, 1, \ldots, 2001\}$ such that

$$y_1 < y_2 < \ldots < y_{1993}.$$

It is not difficult to see that Y has $\binom{2002}{1993}$ elements. We claim that X has the same number of elements and prove it by defining a bijective function $f : X \to Y$. For $(x_n) \in X$, define $(y_n) = f((x_n))$ as follows

$$y_k = x_k + k - 1,$$

for all k, $1 \le k \le 1993$. Clearly,

$$0 \le y_1 < y_2 < \ldots < y_{1993} \le 2001.$$

Conversely, if $(y_n) \in Y$, then $(x_n) = f^{-1}((y_n))$ is determined by

$$x_k = y_k - k + 1,$$

for all k, $1 \le k \le 1993$. A trivial check shows that

$$0 \le x_1 \le x_2 \le \ldots \le x_{1993} \le 9.$$

Thus, the requested number is $\binom{2002}{1993} - 1$, since we have to discard the case when all x_i's are zero (the corresponding number is not a positive integer).

10.3. It is easy to see that the points A, O_1, and O are collinear. Similarly, the points B, O_2, and O are collinear, therefore $\angle O_1OO_2 = \angle AOB$. We will prove that the quadrilateral $ABOD$ is cyclic, implying that $\angle AOB = \angle ADB$, as desired.

Suppose that the tangents at points A and B to $\mathcal{C}$ meet at S (see Figure 10.1). We have

$$\angle OAS = \angle OBS = 90^\circ$$

and since D is the midpoint of KL, we have $\angle ODK = 90^\circ$ as well.

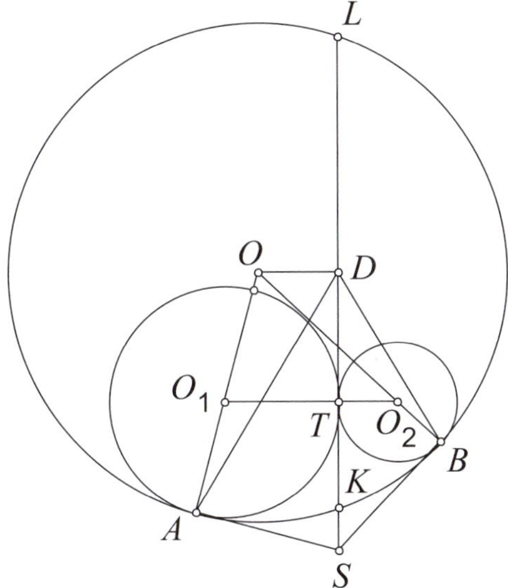

Figure 10.1

If we prove that S lies on the line KL, then it follows that points A, B, and D lie on a circle with diameter OS, thus $ABOD$ is cyclic.

For this purpose, consider an inversion I of pole T. The line KL is invariant under this transformation. The images of the circles $\mathcal{C}_1$ and $\mathcal{C}_2$ are the lines c_1 and c_2, which are perpendicular to the diameters of $\mathcal{C}_1$ and $\mathcal{C}_2$ containing the pole T (see Figure 10.2). Thus, c_1 and c_2 are parallel to KL. The images of the lines AS and BS are two circles a and b, passing through the pole T and tangent to c_1 and c_2, respectively. The second point of intersection between a and b is S', the image of S through the inversion. Clearly, $S'T$ is parallel to c_1 and c_2, therefore S' (and hence S) lies on KL.

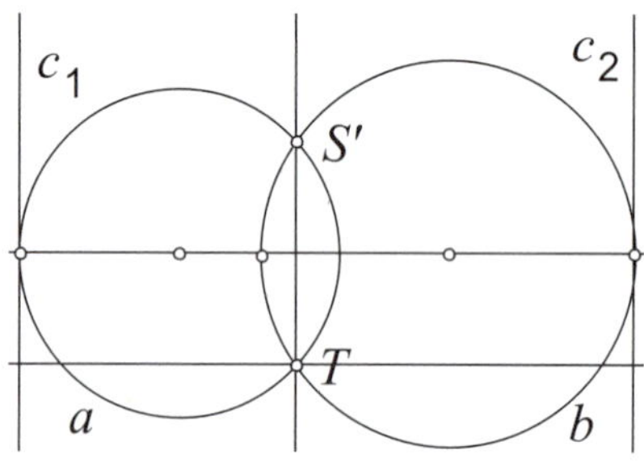

Figure 10.2

Second solution. Let AT and BT meet again the circle $\mathcal{C}$ at the points Q and R, and let AR and BQ intersect at P (see Figure 10.3).

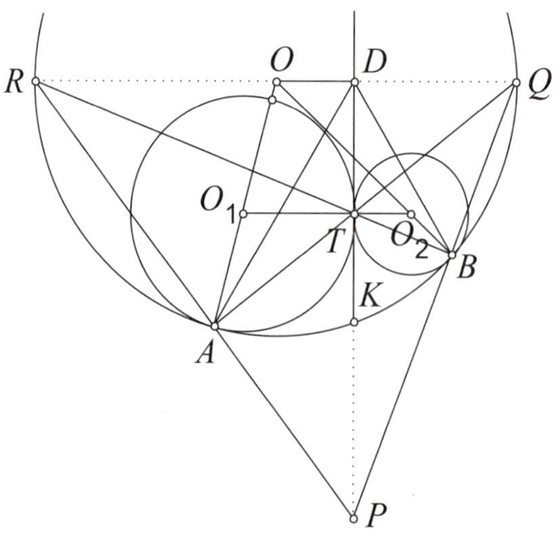

Figure 10.3

Observe that the points Q, O, D, and R are collinear. Indeed, the triangles AO_1T and AOQ are isosceles O_1T is parallel to OQ. In the same way, O_2T is parallel to OR. It follows that O, Q, and R are collinear. But OD and O_1O_2 are both perpendicular to KL, thus D lies on the line QR as well.

Because

$$\angle QAR = \angle QBR = 90^\circ,$$

it results that T is the orthocenter of the triangle PQR and ABD is its orthic triangle. The point O is the midpoint of QR, and hence it belongs to the nine point circle of the triangle PQR, which is the circumcircle of ABD (see Appendix for information about the nine point circle). This implies that the quadrilateral $ABOD$ is cyclic, so that $\angle AOB = \angle ADB$.

10.4. If $m = p$, the equation is

$$\frac{x^p + y^p}{2} = \left(\frac{x+y}{2}\right)^p,$$

and has infinitely many solutions (x, x), where x is an arbitrary positive integer.

Conversely, suppose that $(x, y) \neq (1, 1)$ is a solution. If $x = y$ then $m = p$ and we are done. Assume that $x < y$ and let $d = \gcd(x, y)$. Then $x = ad$ and $y = bd$, for some positive integers a and b, with $\gcd(a, b) = 1$. The equation becomes

$$d^p \frac{a^p + b^p}{2} = d^m \left(\frac{a+b}{2}\right)^m. \tag{1}$$

Jensen's inequality applied to the convex function $f : (0, +\infty) \to \mathbb{R}$, $f(x) = x^p$, gives

$$\frac{a^p + b^p}{2} > \left(\frac{a+b}{2}\right)^p,$$

and it results

$$d^m \left(\frac{a+b}{2}\right)^m > d^p \left(\frac{a+b}{2}\right)^p,$$

yielding $m > p$. Thus, (1) can be written as

$$2^{m-1}(a^p + b^p) = d^{m-p}(a+b)^m. \tag{2}$$

We deduce that $a + b$ is an even number, otherwise $(a+b)^m$ would divide $a^p + b^p$, which is impossible since $m > p$. We can then set $a = c - t$ and $b = c + t$, where c and t are positive integers and $\gcd(c, t) = 1$. Plugging into (2) gives

$$2^{m-1}\left(c^p + \binom{p}{2}c^{p-2}t^2 + \ldots + \binom{p}{p-1}ct^{p-1}\right) = d^{m-p}2^m c^m,$$

or

$$c^{p-1} + \binom{p}{2} c^{p-3} t^2 + \ldots + \binom{p}{p-1} t^{p-1} = 2d^{m-p} c^{m-1}.$$

It follows that c divides pt^{p-1}, and since $\gcd(c, t) = 1$, c divides p. This is possible only if $c = p$ but then the above equality implies that p divides t, which is a contradiction (note that if p is a prime number, then p divides $\binom{p}{k}$, for all k, $1 \leq k \leq p - 1$).

The 11th BMO

The eleventh Balkan Mathematical Olympiad for high-school students was held between May 8th and May 13th, 1994, in Novi Sad, Yugoslavia. The number of the participating countries remained 6: Albania, Bulgaria, Cyprus, Greece, Romania and Yugoslavia.

Problems

11.1. Let P be a point inside the acute angle $\angle xAy$. Construct (by ruler and compass) the points B and C on the halflines Ax and Ay such that the line BC passes through P and the area of the triangle ABC equals AP^2.

(Cyprus)

11.2. Let m be an integer number. Prove that the polynomial

$$P(X) = X^4 - 1994X^3 + (1993 + m)X^2 - 11X + m$$

has at most one integer root.

(Greece)

11.3. Let $n \geq 2$ be a positive integer. Find the maximum value of the expression

$$\sum_{k=1}^{n-1} |a_{k+1} - a_k|$$

where $a_1, a_2, \ldots, a_n$ is a permutation of the numbers $1, 2, \ldots, n$.

(Romania)

11.4. Find the least integer $n > 4$ with the property: there exists a set of n persons such that every two persons knowing each other have no common acquaintances and every two persons not knowing each other have exactly two common acquaintances (we assume that for every two persons A, B, if A knows B, then B knows A).

(Bulgaria)

Solutions

11.1. Let Q be the projection of the point P on the line Ax and R a point on Ax such that PR is parallel to Ay (see Figure 11.1). Observe that since $\angle xAy$ is acute, the point R lies between A and Q.

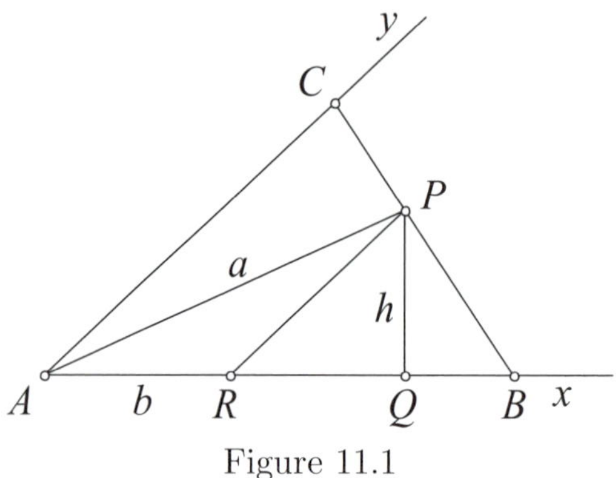

Figure 11.1

Denote $AP = a$, $PQ = h$, $AR = b$, and $RB = x$. The lengths a, b, and h are given, and we have to construct a line segment of length x. The triangles ABC and RBP are similar, hence

$$\frac{[RBP]}{[ABC]} = \left(\frac{RB}{AB}\right)^2 = \left(\frac{x}{x+b}\right)^2.$$

On the other hand, $[RBP] = \frac{1}{2}hx$ and $[ABC] = a^2$. It follows that

$$x^2 - 2\left(\frac{a^2}{h} - b\right)x + b^2 = 0. \qquad (1)$$

We construct now a line segment of length $\frac{a^2}{h}$. Consider a right triangle KLM with legs $KL = h$ and $KM = a$ (see Figure 11.2).

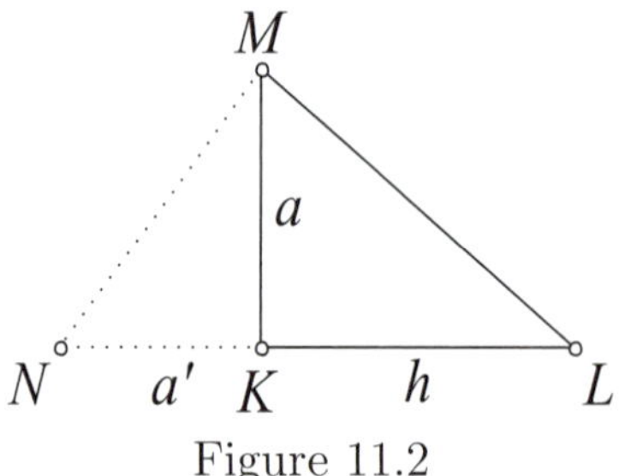

Figure 11.2

The perpendicular erected in M to the line ML intersects the line KL at N. We have

$$MK^2 = NK \cdot KL$$

so

$$a' = NK = \frac{a^2}{h}.$$

Next, it is easy to construct the segment of length $c = a' - b$. This is always possible because

$$\frac{a^2}{h} > b.$$

Indeed, we have

$$a^2 = h^2 + (b+y)^2 > bh,$$

where $y = RQ$. The equation (1) becomes

$$x^2 - 2cx + b^2 = 0.$$

Let x_1 and x_2 be its roots. We have $x_1 + x_2 = 2c$ and $x_1 x_2 = b^2$. Consider a semicircle of diameter $2c$ and draw a parallel to its diameter at a distance equal to b (see Figure 11.3).

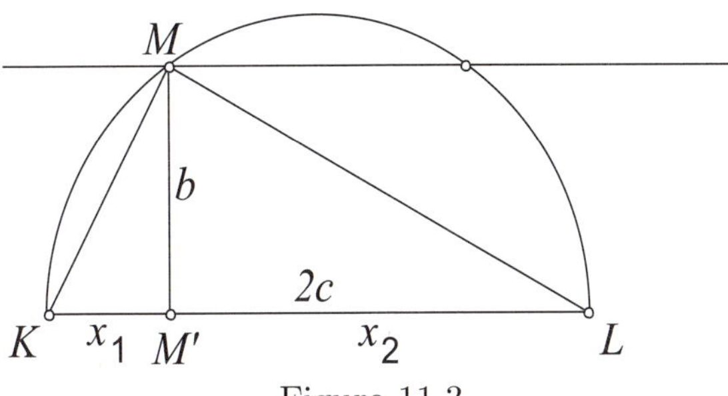

Figure 11.3

If M is one of the points in which the parallel intersects the semicircle and M' is its projection on the diameter KL, then the lengths of the line segments KM' and LM' are the roots x_1 and x_2. This is because $KM' + LM' = KL = 2c$ and the triangle KLM is a right triangle so

$$MM'^2 = KM' \cdot LM'.$$

The problem always has two solutions, because $b < c$. Indeed, this is equivalent to

$$a^2 > 2hb$$

and it follows again from

$$a^2 = h^2 + (b+y)^2.$$

11.2. Suppose that P has at least two integer roots x_1,x_2, and let x_3, x_4 be the other two roots of the polynomial. Then we have (Viète's relations)

$$\begin{aligned} x_1 + x_2 + x_3 + x_4 &= 1994 \\ x_1x_2 + x_1x_3 + x_1x_4 + x_2x_3 + x_2x_4 + x_3x_4 &= 1993 + m \\ x_1x_2x_3 + x_1x_2x_4 + x_1x_3x_4 + x_2x_3x_4 &= 11 \\ x_1x_2x_3x_4 &= m. \end{aligned}$$

Denote $x_1 + x_2 = s_1$, $x_3 + x_4 = s_2$, $x_1x_2 = p_1$, $x_3x_4 = p_2$, and rewrite the above equalities to obtain

$$s_1 + s_2 = 1994 \tag{1}$$

$$p_1 + p_2 + s_1s_2 = 1993 + m \tag{2}$$

$$p_1s_2 + p_2s_1 = 11 \tag{3}$$

$$p_1p_2 = m. \tag{4}$$

Since x_1 and x_2 are integers, s_1 and p_1 are integers as well. Then (1) implies that s_2 is an integer and (2) implies that p_2 is also an integer. We now look at the parity of the numbers s_1, s_2, p_1, and p_2. From (1) we deduce that s_1 and s_2 have the same parity; they cannot be both even because this would contradict (3). Hence s_1 and s_2 are both odd. But then (3) implies that p_1 and p_2 have different parities, so m is even (because of (4)). Therefore, $p_1 + p_2$ and s_1s_2 are odd and m is even. This contradicts (2), so our assumption is false.

We conclude that at most one root can be an integer number.

Second solution. Let f be an integer polynomial,

$$f(X) = a_0 + a_1X + \ldots + a_nX^n,$$

and denote by $\widehat{f}$ the polynomial reduced modulo 2, that is,

$$\widehat{f}(X) = \widehat{a}_0 + \widehat{a}_1X + \ldots + \widehat{a}_nX^n,$$

where $\widehat{a}_i \in \mathbb{Z}_2$, for all i. It is known that if $f = g \cdot h$, where g and h are integer polynomials, then $\widehat{f} = \widehat{g} \cdot \widehat{h}$.

Suppose P has at least two integer roots. Then P factors into $P_1P_2P_3$, where $\deg P_1 = \deg P_2 = 1$ and $\deg P_3 = 2$. Accordingly, we must have $\widehat{P} = \widehat{P}_1\widehat{P}_2\widehat{P}_3$, polynomials in $\mathbb{Z}_2[X]$, with the same degrees as P_1, P_2, and P_3, respectively.

We have

$$\widehat{P}(X) = X^4 + \left(\widehat{1} + \widehat{m}\right)X^2 - X + \widehat{m},$$

so we analyze two cases.

Case 1. If $\widehat{m} = \widehat{0}$, then $\widehat{P}(X) = X^4 + X^2 - X = X\left(X^3 + X - \widehat{1}\right)$. But $Q(X) = X^3 + X - \widehat{1}$ is an irreducible polynomial in $\mathbb{Z}_2[X]$. Indeed, if Q is reducible, one of its factors must have degree 1 and hence Q must have a root in $\mathbb{Z}_2$, which is not the case.

Case 2. If $\widehat{m} = \widehat{1}$, then $\widehat{P}(X) = X^4 - X + \widehat{1}$. Observe that P has no roots in $\mathbb{Z}_2$, hence no factor of degree 1. It follows that P cannot be written as a product of three polynomials in $\mathbb{Z}_2[X]$.

We conclude that P has at most one integer root.

11.3. If $p = (a_1, a_2, \ldots, a_n)$ is a permutation of the numbers $1, 2, \ldots, n$, we denote

$$s(p) = \sum_{k=1}^{n-1} |a_{k+1} - a_k| + |a_1 - a_n|.$$

Observe that

$$s(p) = \sum_{k=1}^{n} \alpha_k (a_{k+1} - a_k) = \sum_{k=1}^{n} (\alpha_{k-1} - \alpha_k) a_k,$$

where $\alpha_k \in \{-1, 1\}$, and we denoted $a_{n+1} = a_1$, $\alpha_0 = \alpha_n$. Rearranging terms, we have

$$s(p) = \sum_{k=1}^{n} \beta_k \cdot k,$$

where $\beta_k \in \{-2, 0, 2\}$ and $\beta_1 + \beta_2 + \ldots + \beta_n = 0$. Thus,

$$s(p) = 2(x_1 + x_2 + \ldots + x_m) - 2(y_1 + y_2 + \ldots + y_m),$$

where $2m$ equals the number of the β_k's different from 0 and the numbers $x_1, \ldots, x_m, y_1, \ldots, y_m$ are distinct elements of $\{1, 2, \ldots, n\}$. We can see that $s(p)$ reaches its maximal value when $m = \left\lfloor \frac{n}{2} \right\rfloor$ and

$$\begin{aligned} \{x_1, \ldots, x_m\} &= \{m+1, m+2, \ldots, n\}, \\ \{y_1, \ldots, y_m\} &= \{1, 2, \ldots, m\}. \end{aligned} \tag{1}$$

Since

$$\sum_{k=1}^{n-1} |a_{k+1} - a_k| = s(p) - |a_1 - a_n| \le s(p) - 1,$$

from all the permutations satisfying (1), for which $s(p)$ is maximal, we must find one with $|a_1 - a_n| = 1$. For n even, $n = 2m$, take $a_{2k-1} = m + k$, $a_{2k} = k$, for $k = 1, 2, \ldots, m$, and for n odd, $n = 2m + 1$, take $a_{2k-1} = m + k + 1$,

$a_{2k} = k$, for $k = 1, 2, \ldots, m$, $a_n = m + 1$. It is not difficult to see that the requested maximal value equals $\frac{n^2-2}{2}$ if n is even, and $\frac{n^2-3}{2}$ if n is odd.

11.4. It is easier to use the language of graph theory. Consider a graph with n vertices; then the first condition means that the graph contains no triangles. The second one reads as follows: for any two vertices which are not connected, there exist exactly two other vertices connected to both.

Let A be an arbitrary vertex and suppose that it is connected to p vertices $A_1, A_2, \ldots, A_p$ (that is, the degree of A equals p). Let N_A be the set of vertices which are not connected to A. We will prove that N_A has $\binom{p}{2}$ elements.

Observe that for every $i \neq j$, A_i and A_j are not connected, otherwise A, A_i, and A_j are the vertices of a triangle. Therefore, for each pair A_i, A_j there exists exactly one vertex from N_A connected to both. Denote by A_{ij} this vertex. If the pairs i, j and h, k are different and $A_{ij} = A_{hk} = B$ then A and B are both connected to at least three vertices, a contradiction. It follows that N_A has at least $\binom{p}{2}$ elements.

If $B \in N_A$, then there must be exactly two vertices connected to both A and B, say A_i and A_j. For another vertex $C \in N_A$ different from B, there exist A_h and A_k connected to C. The pairs i, j and h, k must be different, otherwise the vertices A, B, and C are all connected to the same two vertices. It follows that N_A has at most $\binom{p}{2}$ elements.

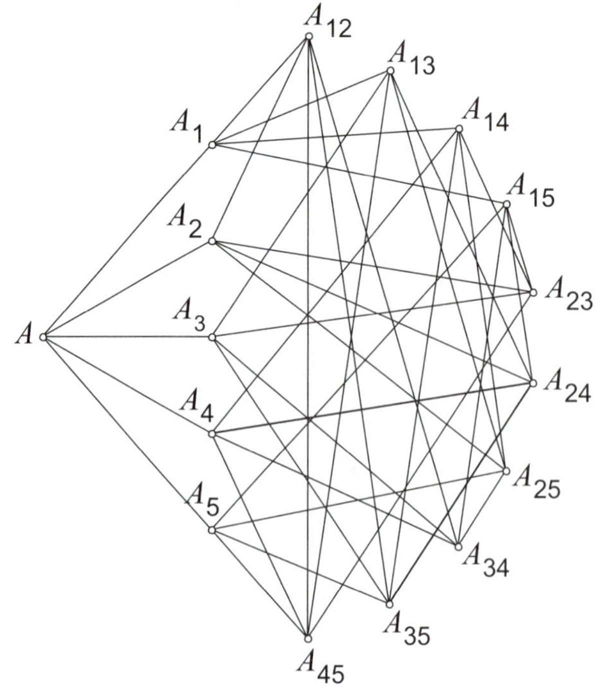

Figure 11.4

We conclude that N_A has exactly $\binom{p}{2}$ elements, hence the total number of vertices is

$$n = 1 + p + \binom{p}{2}.$$

Observe that A_1 is connected to A and to $p-1$ points A_{1j}, thus its degree equals p. Since A was an arbitrary vertex, we find that every two connected vertices have the same degree, hence all vertices have the same degree p.

Since $n > 4$, we have $p > 2$. For $p = 3$ it results $n = 7$. We have the vertices A, A_1, A_2, A_3, A_{12}, A_{13}, and A_{23}. The vertex A_1 is connected to A_{12} and A_{13}, etc. But A_{12} cannot be connected to neither A_{13} nor A_{23}, otherwise a triangle is formed ($A_1A_{12}A_{13}$ in the first case, $A_2A_{12}A_{23}$ in the second), so its degree equals 2, a contradiction. Similarly, the case $p = 4$ (hence $n = 11$) leads to a contradiction.

For $p = 5$ we have $n = 16$ and there exists a graph with 16 vertices and the requested properties (see Figure 11.4).

The 12th BMO

The twelfth Balkan Mathematical Olympiad for high-school students was held between May 8th and May 13th , 1995, in Plovdiv, Bulgaria. The number of the participating countries increased to 7: Albania, Bulgaria, Cyprus, Former Yugoslav Republic of Macedonia, Greece, Romania and Yugoslavia.

Problems

12.1. Find the value of the expression $(\ldots(((2\star 3)\star)4)\star\ldots)\star 1995$, where

$$x\star y=\frac{x+y}{1+xy},$$

for all positive numbers x,y.

(F.Y.R. Macedonia)

12.2. The circles C_1 and C_2 with centers O_1, O_2 and radii r_1, r_2, respectively, intersect at A and B such that $\angle O_1AO_2=90°$. The line O_1O_2 intersects C_1 at C,D and C_2 at E,F, such that E lies between C and D and D lies between E and F. The line BE intersects the second time C_1 at K and intersects AC at M. The line BD intersects the second time C_2 at L and intersects AF at N. Show that

$$\frac{r_2}{r_1}=\frac{KE}{KM}\cdot\frac{LN}{LD}.$$

(Greece)

12.3. Let a and b be positive integers having the same parity, such that $a>b$. Prove that the roots of the equation

$$x^2-\left(a^2-a+1\right)\left(x-b^2-1\right)-\left(b^2+1\right)^2=0$$

are positive integers, none of which is a perfect square.

(Albania)

12.4. Let n be a positive integer and let S be the set of all points (x, y), where x and y are positive integers and $x, y \leq n$. Let T be the set of all squares whose vertices belong to S. Denote by a_k $(k \geq 0)$ the number of pairs of points in S which are the vertices of exactly k squares from T. Prove that $a_0 = a_2 + 2a_3$.

(Yugoslavia)

Solutions

12.1. Consider the function $f : (0, +\infty) \to (-1, 1)$, $f(x) = \dfrac{1+x}{1-x}$ and observe that

$$f(x \star y) = \frac{1 + \frac{x+y}{1+xy}}{1 - \frac{x+y}{1+xy}} = \frac{(1+x)(1+y)}{(1-x)(1-y)} = f(x) f(y).$$

Therefore, if

$$a = (\ldots(((2 \star 3)\star)4) \star \ldots) \star 1995,$$

then

$$f(a) = f(2) \cdot f(3) \cdot \ldots \cdot f(1995)$$
$$= \frac{3}{-1} \cdot \frac{4}{-2} \cdot \frac{5}{-3} \cdot \ldots \cdot \frac{1994}{-1992} \cdot \frac{1995}{-1993} \cdot \frac{1996}{-1994} = \frac{1995 \cdot 1996}{2}.$$

The equation

$$f(a) = \frac{1+a}{1-a} = \frac{1995 \cdot 1996}{2} = 1991010$$

has the solution $a = \dfrac{1991\,009}{1991\,011}$.

12.2. Observe that the points C, A, and L are collinear. Indeed, we have

$$\angle CAL = \angle CAO_1 + 90^\circ + \angle O_2AL.$$

But

$$\angle O_2AL = 90^\circ - \frac{\angle AO_2L}{2} = 90^\circ - \angle ABL = 90^\circ - \angle ABD = 90^\circ - \angle ACO_1.$$

And since $\angle CAO_1 = \angle ACO_1$, we obtain that $\angle CAL = 180^\circ$, that is, the points C, A, and L are collinear. Similarly, we have that K, A, and F are collinear as well.

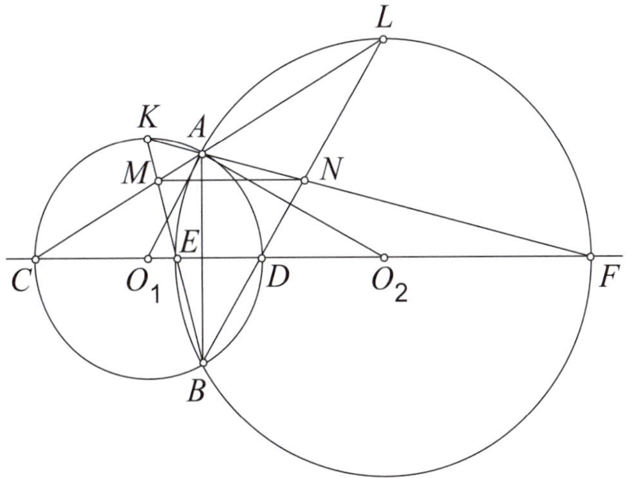

Figure 12.1

Applying Menelaus' theorem to the triangle CME and the line KF yields

$$\frac{KE}{KM} \cdot \frac{AM}{AC} \cdot \frac{FC}{FE} = 1.$$

The same theorem applied to FND and CL gives

$$\frac{LN}{LD} \cdot \frac{AF}{AN} \cdot \frac{CD}{CF} = 1.$$

We deduce that

$$\frac{KE}{KM} \cdot \frac{LN}{LD} = \frac{AC}{AM} \cdot \frac{AN}{AF} \cdot \frac{r_2}{r_1},$$

since $FE = 2r_2$ and $CD = 2r_1$. Therefore, it suffices to prove that

$$\frac{AC}{AM} \cdot \frac{AN}{AF} = 1,$$

or, equivalently,

$$\frac{AC}{AM} = \frac{AF}{AN},$$

that is, MN is parallel to CF.

We have

$$\begin{aligned}
\angle MAN &= \angle CAO_1 + 90^\circ + \angle O_2AF \\
&= 90^\circ - \frac{\angle CO_1A}{2} + 90^\circ + 90^\circ - \frac{\angle AO_2F}{2} \\
&= 270^\circ - \frac{1}{2}(180^\circ - \angle AO_1O_2 + 180^\circ - \angle AO_2O_1) \\
&= 270^\circ - \frac{1}{2}(360^\circ - 90^\circ) = 135^\circ.
\end{aligned}$$

On the other hand,

$$\angle MBN = \angle EBA + \angle DBA = \frac{\angle AO_2O_1}{2} + \frac{\angle AO_1O_2}{2} = 45^\circ.$$

It follows that the quadrilateral $AMBN$ is cyclic, therefore

$$\angle AMN = \angle ABN = \angle ABD = \angle ACD,$$

which implies that MN is parallel to CF, as desired.

12.3. The solutions of the quadratic equation are

$$x_1 = b^2 + 1 \text{ and } x_2 = a^2 - a - b^2.$$

It is easy to see that $b^2 + 1$ cannot be a square, so we have to prove that the equation

$$a^2 - a - b^2 = c^2 \tag{1}$$

has no solution in positive integers with a and b having the same parity.

Suppose the contrary and let

$$u = \frac{a+b}{2}, \quad v = \frac{a-b}{2}.$$

Since a and b have the same parity and $a > b$, u and v are positive integers. We obtain

$$a = u + v, \quad b = u - v$$

and replacing in (1) yields

$$4uv - u - v = c^2,$$

or, equivalently,

$$(4u - 1)(4v - 1) = 4c^2 + 1.$$

Because $4u - 1 \equiv -1 \pmod 4$ there exists at least one prime number p dividing $4u - 1$ such that $p \equiv -1 \pmod 4$ (in fact, the number of such prime numbers is odd). We have then

$$(2c)^2 \equiv -1 \pmod p,$$

and since $\frac{p-1}{2}$ is an odd number, the above congruence implies

$$(2c)^{p-1} \equiv -1 \pmod p.$$

Clearly, $\gcd(2c, p) = 1$, so Fermat's little theorem yields

$$(2c)^{p-1} \equiv 1 \pmod p.$$

It follows that

$$1 \equiv -1 \pmod{p},$$

an obvious contradiction.

Observation. The equation (1) has infinitely many solutions in positive integers with a and b of different parities. A family of solutions is given by

$$a = k^2 + 1, \ b = k^2, \ c = k,$$

where k is an arbitrary integer.

12.4. The total number of pairs of points from S is

$$\binom{n^2}{2} = \frac{n^2\left(n^2 - 1\right)}{2}.$$

Observe that there exist pairs of points in S which are not the vertices of any square from T (e.g. the pair A, B in Figure 12.2). Also, some pairs are the vertices of exactly one (like C, D), two (like E, F) or three squares from T (like G, H).

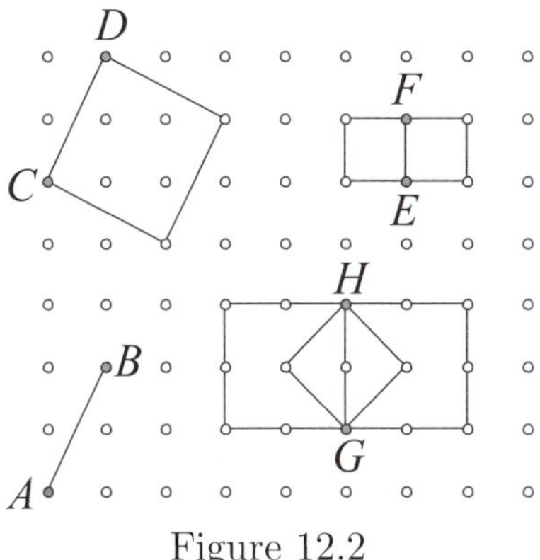

Figure 12.2

However, we cannot find two points in S which are the vertices of more than three squares. Indeed, a pair of points can represent the vertices of a side of a square (and there are at most two squares in T sharing a given side) or of a diagonal (for at most one square).

We conclude that the total number of pairs of points in S equals

$$a_0 + a_1 + a_2 + a_3.$$

Let t be the number of elements of the set T. Counting for each square the pairs of points that are among its vertices, we find that their total number is

$$a_1 + 2a_2 + 3a_3.$$

On the other hand, this number must be equal to $6t$, since in each square there are 6 pairs of points among its vertices. If we prove that

$$6t = \frac{n^2\left(n^2-1\right)}{2},$$

then it will follow that

$$a_0 + a_1 + a_2 + a_3 = a_1 + 2a_2 + 3a_3,$$

and hence

$$a_0 = a_2 + 2a_3,$$

the requested equality.

In order to estimate t, we observe that each square in T can be "inscribed" in a square whose sides are parallel to the axis (for instance, in Figure 12.3, $ABCD$ is "inscribed" in $XYZT$). If the side length of the latter equals k, $1 \le k \le n$, then there are exactly k squares "inscribed" in it, including itself.

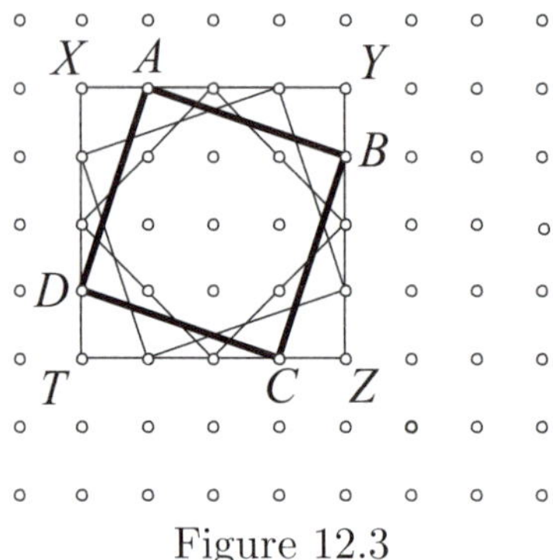

Figure 12.3

It is easy to see that the number of squares from T with side length k and sides parallel to the axis is $(n-k)^2$, therefore the total number of elements of T equals

$$\begin{aligned}\sum_{k=1}^{n} k\left(n-k\right)^2 &= n\sum_{k=1}^{n}\left(n-k\right)^2 - \sum_{k=1}^{n}\left(n-k\right)^3 \\ &= n\cdot\frac{n\left(n+1\right)\left(2n+1\right)}{6} - \frac{n^2\left(n+1\right)^2}{4} \\ &= \frac{n^2\left(n^2-1\right)}{12}.\end{aligned}$$

Thus, we found that

$$6t = \frac{n^2\left(n^2-1\right)}{2},$$

hence the desired conclusion.

The 13th BMO

The thirteenth Balkan Mathematical Olympiad for high-school students was held between April 30th and May 5th , 1996, in Bacău, Romania. The number of the participating countries increased to 9: Albania, Bulgaria, Cyprus, Former Yugoslav Republic of Macedonia, Greece, Republic of Moldova, Romania, Turkey and Yugoslavia.

Problems

13.1. Let O and G be the circumcenter and the centroid of a triangle ABC. If R is its circumradius and r its inradius, show that

$$OG \leq \sqrt{R(R-2r)}.$$

(Greece)

13.2. Let $p > 5$ be a prime number and let

$$X = \left\{p - n^2 \mid n \text{ is a positive integer and } n^2 < p\right\}.$$

Prove that X contains two different elements x, y such that $x \neq 1$ and x divides y.

(Albania)

13.3. Let $ABCDE$ be a convex pentagon. Denote by M, N, P, Q, and R the midpoints of the sides AB, BC, CD, DE, and EA, respectively. If the segments AP, BQ, CR, and DM have a common point, prove that this point also belongs to the segment EN.

(Yugoslavia)

13.4. Show that there exists a subset A of the set $\left\{1, 2, \ldots, 2^{1996} - 1\right\}$ having the following properties:

a) 1 and $2^{1996} - 1$ belong to A;

b) every element of A except 1 is the sum of two (not necessarily distinct) elements of A;

c) the number of elements of A does not exceed 2012.

(Romania)

Solutions

13.1. It is known (see Appendix, Leibniz Relation) that in any triangle the following identity holds

$$OG^2 = R^2 - \frac{1}{9}\left(a^2 + b^2 + c^2\right).$$

Therefore, the requested inequality is equivalent to

$$a^2 + b^2 + c^2 \geq 18Rr.$$

If K and s denote the triangle's area and semiperimeter, we know that

$$K = sr = \frac{abc}{4R},$$

and hence

$$Rr = \frac{abc}{2\left(a + b + c\right)}.$$

It follows that we have to prove the inequality

$$\left(a + b + c\right)\left(a^2 + b^2 + c^2\right) \geq 9abc,$$

which follows by multiplying the obvious AM-GM inequalities

$$a + b + c \geq 3\sqrt[3]{abc}$$

and

$$a^2 + b^2 + c^2 \geq 3\sqrt[3]{a^2b^2c^2}.$$

It is easy to see that the equality occurs if and only if $a = b = c$, i.e. the triangle ABC is equilateral.
Observation. It is known that $OI^2 = R\left(R - 2r\right)$, where I is the incenter of the triangle ABC (Euler triangle formula- see Appendix). Therefore, the problem asks to prove that $OG \leq OI$.

13.2. Suppose that $1 \in X$. Then $p = n^2 + 1$ for some even integer n. Let $x = 2n$ and $y = n^2$. Clearly, x divides y, so we have to prove that both x and y are different elements of X. It is easy to see that

$$2n = n^2 + 1 - \left(n - 1\right)^2 = p - \left(n - 1\right)^2,$$

hence $x \in X$. On the other hand,

$$y = n^2 = p - 1 = p - 1^2,$$

thus $y \in X$ as well. If $x = y$, then $2n = n^2$, hence $n = 2$ and $p = 5$, a contradiction.

Now, suppose that $1 \notin X$ and denote by n the greatest positive integer such that $n^2 < p$ (in fact, $n = \lfloor \sqrt{p} \rfloor$). Let $x = p - n^2$ and $y = p - (x - n)^2$. Observe that

$$y = p - n^2 + 2nx - x^2 = x + 2nx - x^2 = x\,(1 + 2n - x)\,,$$

hence x divides y. Clearly, x belongs to X. In order to show that y is also an element of X different from x, it is sufficient to prove that

$$1 \le |x - n| < n.$$

Indeed, if $|x - n| = 0$, then $p = n^2 + n = n\,(n+1)$, which is not a prime number. For the second inequality, observe that the definition of n implies

$$p \le n^2 + 2n + 1$$

and since p is a prime,

$$p < n^2 + 2n.$$

Therefore $|x - n| = \left|p - n^2 - n\right| < n$, as requested.

13.3. We will use the following result.
Lemma. Let X' be the midpoint of the side YZ of the triangle XYZ. The point O belongs to the median XX' if and only if O is an interior point of the triangle and $[OXY] = [OXZ]$.
Proof. Let O be a point in the interior of the triangle and suppose AO intersects YZ at X'. Project Y and Z on the line AO to the points Y', Z, respectively. Triangles $YY'X'$ and $ZZ'X'$ are similar. Since the triangles OXZ and OXY have the common side AO, then $[OXY] = [OXZ]$ if and only if $YY' = ZZ'$, that is, if and only if $YX' = ZX'$ (see Figure 13.1).

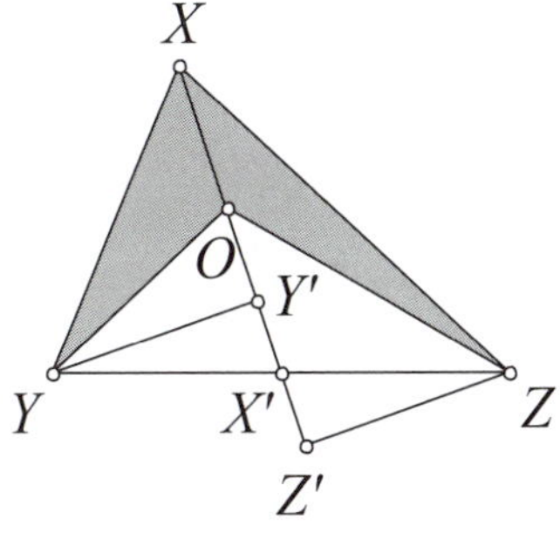

Figure 13.1

Returning to our problem, let O be the common point of the segments AP, BQ, CR, and DM. Observe that from the hypothesis and the lemma we have

$$[BOE] = [BOD] = [AOD] = [AOC] = [COE].$$

It is not difficult to see that O lies in the interior of BEC and since $[BOE] = [COE]$, the line EO passes through the midpoint N of the segment BC (see Figure 13.2).

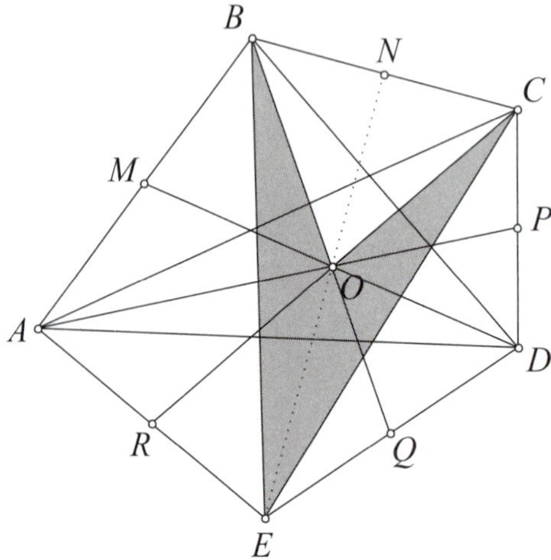

Figure 13.2

Second solution. Let O be the origin of the complex plane and let $a, b, c, \ldots$ be the affixes of the points $A, B, C, \ldots$. Observe that the distinct points O, X, Y are collinear if and only if there exists a real number λ such that $x = \lambda y$, or, equivalently, $\frac{x}{y}$ is a real number. But this happens if and only if

$$\frac{x}{y} = \overline{\left(\frac{x}{y}\right)} = \frac{\overline{x}}{\overline{y}}$$

or

$$x\overline{y} = \overline{x}y.$$

Since P is the midpoint of CD, we have

$$p = \frac{c+d}{2}$$

and because O, A, and P are collinear, we have

$$a\left(\overline{c} + \overline{d}\right) = \overline{a}\,(c+d).$$

Similarly,

$$b\left(\overline{d}+\overline{e}\right)=\overline{b}\left(d+e\right),$$
$$c\left(\overline{e}+\overline{a}\right)=\overline{c}\left(e+a\right),$$
$$d(\overline{a}+\overline{b})=\overline{d}\left(a+b\right).$$

Adding up these equalities yields

$$\overline{e}\left(b+c\right)=e\left(\overline{b}+\overline{c}\right),$$

and it follows that O, E and N are collinear as well.

13.4. Let $f(n)$ be the least number of elements of a set $A\subset\{1,2,\ldots,2^n-1\}$ satisfying conditions a) and b). For instance, $f(3)\leq 5$, since the set

$$A=\{1,2,3,6,7\}\subset\{1,2,3,4,5,6,7\}$$

satisfies the required conditions. Thus, we have to prove that $f(1996)\leq 2012$.

First, observe that

$$f(n+1)\leq f(n)+2. \tag{1}$$

Indeed, if $A\subset\{1,2,\ldots,2^n-1\}$ satisfies a) and b), so does

$$B=A\cup\left\{2^{n+1}-2,2^{n+1}-1\right\},$$

since

$$2^{n+1}-2=2\left(2^n-1\right)$$

and

$$2^{n+1}-1=1+(2^{n+1}-2).$$

Also, notice that

$$f(2n)\leq f(n)+n+1. \tag{2}$$

This is because

$$B=A\cup\left\{2\left(2^n-1\right),2^2\left(2^n-1\right),\ldots,2^n\left(2^n-1\right),2^{2n}-1\right\}$$

satisfies a) and b):

$$2^{k+1}\left(2^n-1\right)=2\cdot 2^k\left(2^n-1\right),$$

for $k=0,1,\ldots,n-1$ and

$$2^{2n}-1=2^n\left(2^n-1\right)+\left(2^n-1\right).$$

Now, we can obtain an upper estimate of $f(m)$ using successively either (1) or (2), according to the parity of the argument of f. To shorten the computation, observe that (1) and (2) imply

$$f(2n+1) \le f(n) + n + 3,$$

and a simple induction shows that

$$f\left(2^k(2n+1)\right) \le f(n) + k + m + 3 + (2n+1)\left(2^k - 1\right).$$

Since $1996 = 2^2 \cdot 499 = 2^2(2 \cdot 249 + 1)$, we have

$$f(1996) \le f(249) + 1751.$$

In a similar way, we obtain

$$\begin{aligned} f(249) &\le f(124) + 127, \\ f(124) &\le f(15) + 113, \\ f(15) &\le f(7) + 10, \\ f(7) &\le f(3) + 6, \end{aligned}$$

and, as we saw, $f(3) \le 5$. Adding up yields $f(1996) \le 2012$, as needed.
Observation. The result can be improved[3]. Indeed, we claim that if we impose to the set A the additional condition $7 \in A$, then the following inequality holds:

$$f(n+3) \le f(n) + 4. \tag{3}$$

To prove this, observe that if $A \subset \{1, 2, \ldots, 2^n - 1\}$ satisfies a) and b) and $7 \in A$, so does

$$B = A \cup \left\{2^{n+1} - 2, 2^{n+2} - 4, 2^{n+3} - 8, 2^{n+3} - 1\right\}.$$

We now have

$$\begin{aligned} f(1996) &\le f(499) + 1499 \\ f(499) &\le f(496) + 4 \\ f(496) &\le f(31) + 469 \\ f(31) &\le f(28) + 4 \\ f(28) &\le f(7) + 23, \end{aligned}$$

and, finally,

$$f(7) \le 11$$

as before. Adding up, we obtain $f(1996) \le 2010$.

[3]Thanks to Gheorghe Eckstein for showing us his solution.

The 14th BMO

The fourteenth Balkan Mathematical Olympiad for high-school students was held between April 29th and May 4th, 1997, in Kalampaka, Greece. The number of the participating countries remained 9: Albania, Bulgaria, Cyprus, Former Yugoslav Republic of Macedonia, Greece, Republic of Moldova, Romania, Turkey and Yugoslavia. The neighborings rocks of the mountains and wonderful monasteries of Meteora gave to the competition a particular beauty.

Problems

14.1. Let O be an interior point of the convex quadrilateral $ABCD$, satisfying the condition

$$OA^2 + OB^2 + OC^2 + OD^2 = 2\,[ABCD]\,.$$

Show that $ABCD$ is a square and O is its center.

(Yugoslavia)

14.2. Let $m, n \geq 2$ be integer numbers, let S be a set with n elements and $A_1, A_2, \ldots, A_m$ be given subsets of S satisfying the following condition: for any two distinct elements x, y of S there is a subset A_i such that either $x \in A_i$ and $y \notin A_i$ or $x \notin A_i$ and $y \in A_i$.

Prove that $2^m \geq n$.

(Yugoslavia)

14.3. Consider three circles $C_1, C_2,$ and Γ such that C_1, C_2 are internally tangent to Γ at the points B, C, respectively, and are also externally tangent at a point D. Let A be one of the points at which the common internal tangent to C_1 and C_2 at D meets the circle Γ. Moreover, let K, L be the points of intersection of AB, AC with the circles C_1, C_2, respectively, and M, N be the points of intersection of BC with the same circles.

Prove that the lines AD, KM, and LN pass through the same point P.

(Greece)

14.4. Find all functions $f : \mathbb{R} \to \mathbb{R}$ such that

$$f(xf(x) + f(y)) = (f(x))^2 + y,$$

for all real x and y.

(Bulgaria)

Solutions

14.1. We have

$$[ABCD] = [OAB] + [OBC] + [OCD] + [OAD].$$

But

$$[OAB] = \frac{1}{2} OA \cdot OB \sin \angle AOB \leq \frac{1}{2} OA \cdot OB,$$

and, using the similar inequalities, we obtain

$$2[ABCD] \leq OA \cdot OB + OB \cdot OC + OC \cdot OD + OD \cdot OA. \tag{1}$$

On the other hand, Cauchy Schwartz inequality yields

$$\begin{aligned} &OA \cdot OB + OB \cdot OC + OC \cdot OD + OD \cdot OA \\ &\leq OA^2 + OB^2 + OC^2 + OD^2 \end{aligned} \tag{2}$$

and equality holds if and only if

$$\frac{OA}{OB} = \frac{OB}{OC} = \frac{OC}{OD} = \frac{OD}{OA},$$

that is, if and only if $OA = OB = OC = OD$.

The hypothesis of the problem claims that the equality holds in both inequalities (1) and (2). Therefore $OA = OB = OC = OD$ and $\sin \angle AOB = \sin \angle BOC = \sin \angle COD = \sin \angle DOA = 1$. This proves that $ABCD$ is a square and O is its center.

Observation. Even simpler inequalities can be used in solving the problem:

$$\begin{aligned} 2[ABCD] &= \sum OA \cdot OB \sin \angle AOB \\ &\leq \sum OA \cdot OB \leq \sum \frac{OA^2 + OB^2}{2} = \sum OA^2. \end{aligned}$$

We deduce $\sin \angle AOB = 1$, $OA = OB$ and the similar equalities. The conclusion follows immediately.

14.2. Let $S = \{x_1, x_2, \ldots, x_n\}$ and let $A = (a_{ij})_{\substack{1\le i\le m\\ 1\le j\le n}}$ be the incidence matrix of the family of subsets $A_1, A_2, \ldots, A_m$, that is,

$$a_{ij} = \begin{cases} 1, & x_j \in A_i \\ 0, & x_j \in \overline{A_i}. \end{cases}$$

The given condition means that the columns of A are all distinct. This condition can be read as follows: the function which maps the element x_j in the column C_j of A is one to one. The number of all columns is 2^m and the size of S is n, hence $n \le 2^m$.

Second solution. For any integer k, $0 \le k \le m$ and any ordered sequence $1 \le i_1 < i_2 < \ldots < i_k \le m$, construct the set

$$S_{i_1 i_2 \ldots i_k} = A_1 \cap \ldots \cap \overline{A_{i_1}} \cap \ldots \cap \overline{A_{i_k}} \cap \ldots \cap A_m,$$

and include

$$S_\emptyset = A_1 \cap \ldots \cap A_m$$

and

$$S_{12\ldots m} = \overline{A_1} \cap \ldots \cap \overline{A_m}$$

The number of these sets is 2^m. By the hypothesis, each element of S belongs to precisely one of the sets $S_{i_1 i_2 \ldots i_k}$. This proves that $n \le 2^m$.

14.3. Let P be the point of intersection between KM and LN. We will show that the quadrilateral $ALPK$ is a parallelogram and AD is one of its diagonals (see Figure 14.1).

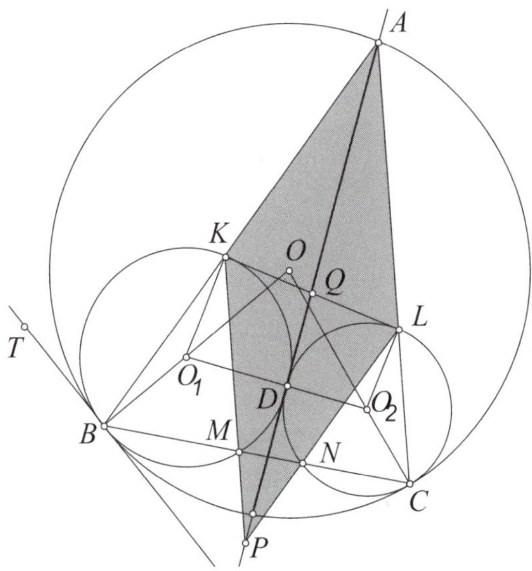

Figure 14.1

The key point is to prove that KL is the common external tangent of the circles C_1 and C_2. For this, we prove that

$$\angle O_1KL = \angle O_2LK = 90^\circ,$$

where O_1 and O_2 are the centers of C_1 and C_2.

By the power of a point theorem we have

$$AK \cdot AB = AD^2 = AL \cdot AC.$$

Therefore

$$\frac{AK}{AC} = \frac{AL}{AB},$$

hence the triangles AKL and ACB are similar.
It follows that $\angle AKL = \angle ACB$ and $\angle ALK = \angle ABC$. Since Γ and C_1 have the common tangent BT in point B we have

$$\angle ABT = \frac{1}{2}\text{arc}AB = \frac{1}{2}\text{arc}BK$$

and then $\angle BKO_1 = \angle BAO$. Therefore O_1K and OA are parallel lines and

$$\angle BKO_1 = 90^\circ - \frac{1}{2}\angle BOA = 90^\circ - \angle BCA = 90^\circ - \angle AKL.$$

This shows that O_1K is orthogonal to KL. In the same way, O_2L is orthogonal to KL, thus proving that KL is the common tangent. Also, if Q is the intersection between AD and KL, we have $KQ = QD = QL$.

We prove now that MK is parallel to AL. In the triangles BMK and BCA we have

$$\angle BMK = \frac{1}{2}\angle BO_1K = \frac{1}{2}\angle BOA = \angle BCA$$

and $\angle KBM = \angle ABC$. Therefore $\angle BKM = \angle BAC$; this proves that MK is parallel to AL. Similarly, NL is parallel to AK so it follows that $ALPK$ is a parallelogram.

Moreover, the line AD passes through the midpoint Q of the diagonal KL, hence it is a diagonal of $ALPK$ as well.

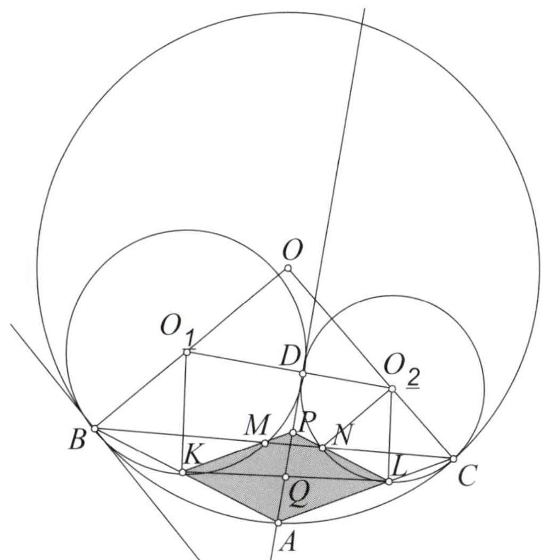

Figure 14.2

Similar considerations prove the result when the point A is taken as the second intersection point between Γ and the tangent in D (see Figure 14.2).

14.4. Plugging $x = 0$ into the given condition yields

$$f(f(y)) = f(0)^2 + y, \tag{3}$$

for all real y. Set $y = y_0 = -f(0)^2$ in the above equality to obtain

$$f(f(y_0)) = 0.$$

Therefore, there exists x_0 such that $f(x_0) = 0$.

For $x = x_0$ and arbitrary y we obtain

$$f(f(y)) = y,$$

which shows that f is a bijective function and $f^{-1} = f$; moreover, comparing to (3) yields $f(0) = 0$.

Setting $y = 0$ in the initial condition we get for any real x

$$f(xf(x)) = f(x)^2.$$

Since $f^{-1} = f$, we can simultaneously replace x by $f(x)$ and $f(x)$ by x in the above equality yielding

$$f(f(x)x) = x^2,$$

and hence

$$f(x)^2 = x^2,$$

for all real x. This means that for each real x, $f(x)$ equals either x or $-x$.

Finally, we prove that only two functions satisfy the given condition, namely $f_1(x) = x$, for all x, and $f_2(x) = -x$, for all x. Indeed, define the sets

$$A = \{x \in \mathbb{R}^* | f(x) = -x\}$$
$$B = \{x \in \mathbb{R}^* | f(x) = x\}.$$

We have $A \cup B = \mathbb{R}^*$. If both sets are nonempty, let $a \in A$ and $b \in B$. Setting $x = a$ and $y = b$ in the given condition, we obtain

$$f\left(-a^2 + b\right) = a^2 + b.$$

But $f\left(-a^2 + b\right)$ equals either $-a^2 + b$ or $-\left(-a^2 + b\right) = a^2 - b$. It follows that either $a = 0$ or $b = 0$, a contradiction.

It is easy to see that f_1 and f_2 indeed satisfy the given condition.

The 15th BMO

The fifteenth Balkan Mathematical Olympiad for high-school students was held between May 3rd and May 9th, 1998, in Nicosia, Cyprus. The number of the participating countries was 8: Albania, Bulgaria, Cyprus, Former Yugoslav Republic of Macedonia, Greece, Republic of Moldova, Romania and Yugoslavia.

Problems

15.1. Find the number of different terms of the finite sequence $\left\lfloor \frac{k^2}{1998} \right\rfloor$, where $k = 1, 2, \ldots, 1997$. ($\lfloor x \rfloor$ denotes the integer part of the real number x.)

(Greece)

15.2. Let $n \geq 2$ be a positive integer and let $0 < a_1 < a_2 < \ldots < a_{2n+1}$ be real numbers. Prove the inequality

$$\sqrt[n]{a_1} - \sqrt[n]{a_2} + \sqrt[n]{a_3} - \ldots + \sqrt[n]{a_{2n+1}} < \sqrt[n]{a_1 - a_2 + a_3 - \ldots + a_{2n+1}}.$$

(Romania)

15.3. Let T be a point inside the triangle ABC and let S be the set of all points except T which are inside or on the border of the triangle ABC. Prove that S can be represented as a disjoint reunion of closed segments. (A closed segment contains both of its ends.)

(Serbia-Montenegro)

15.4. Prove that the equation

$$y^2 = x^5 - 4$$

has no integer solutions.

(Bulgaria)

Solutions

15.1. Let $a_k = \dfrac{k^2}{1998}$ and observe that

$$\frac{(k+1)^2}{1998} - \frac{k^2}{1998} = \frac{2k+1}{1998},$$

therefore, for $k \geq 999$ we have $a_{k+1} - a_k > 1$, implying that the 999 integer parts of the numbers $a_{999}, a_{1000}, \ldots, a_{1997}$ are all different.

On the other hand, since for $k \leq 998$ we have $a_{k+1} - a_k < 1$, it follows that each of the integer parts of the numbers $a_1, a_2, \ldots, a_{998}$ appears at least once in the sequence $0, 1, \ldots, 498 = \left\lfloor \dfrac{998^2}{1997} \right\rfloor$. Thus, the requested number is $999 + 499 = 1498$.

15.2. We prove a more general statement.
Lemma. Let $I \subset \mathbb{R}$ be an arbitrary interval and $f : I \to \mathbb{R}$ a function with the following property: for any numbers $a, b, c \in I$, such that $a < b < c$, it follows

$$f(a) - f(b) + f(c) < f(a - b + c).$$

Then for any numbers $a_1 < a_2 < \ldots < a_{2k+1}$ from I, we have

$$f(a_1) - f(a_2) + f(a_3) - \ldots + f(a_{2k+1}) < f(a_1 - a_2 + \ldots + a_{2k+1}).$$

Proof. First, we must check that $a - b + c \in I$. This is obvious, since $a < a - b + c < c$. Next, we use induction on k. Suppose the statement is true for $k - 1$. Then

$$f(a_1) - f(a_2) + f(a_3) - \ldots + f(a_{2k+1}) <$$
$$< f(a_1 - a_2 + \ldots + a_{2k-1}) - f(a_{2k}) + f(a_{2k+1}).$$

If we denote $a = a_1 - a_2 + \ldots + a_{2k-1}$, $b = a_{2k}$ and $c = a_{2k+1}$, we observe that $a < b < c$, therefore

$$f(a_1 - a_2 + \ldots + a_{2k-1}) - f(a_{2k}) + f(a_{2k+1}) < f(a_1 - a_2 + \ldots + a_{2k+1}),$$

and the lemma is proved.

Returning to our problem, let us check that the function $f : (0, +\infty) \to \mathbb{R}$, $f(x) = \sqrt[n]{x}$ has the property from the lemma. Let $0 < a < b < c$. We have to prove that

$$\sqrt[n]{a} - \sqrt[n]{b} + \sqrt[n]{c} < \sqrt[n]{a - b + c}.$$

Write this under the form

$$\sqrt[n]{b} - \sqrt[n]{a} > \sqrt[n]{c} - \sqrt[n]{a - b + c}$$

and take the conjugate in both sides. We obtain

$$\frac{b-a}{\sqrt[n]{b^{n-1}}+\ldots+\sqrt[n]{a^{n-1}}}>\frac{c-(a-b+c)}{\sqrt[n]{c^{n-1}}+\ldots+\sqrt[n]{(a-b+c)^{n-1}}},$$

which is true, since $b<c$ and $a<a-b+c$.

Using the lemma, we deduce that if $0<a_1<a_2<\ldots<a_{2k+1}$, then

$$\sqrt[n]{a_1}-\sqrt[n]{a_2}+\sqrt[n]{a_3}-\ldots+\sqrt[n]{a_{2k+1}}<\sqrt[n]{a_1-a_2+a_3-\ldots+a_{2k+1}}.$$

Taking $k=n$ yields the desired result.

Observation. It is not difficult to see that all strictly concave down functions have the property from the lemma. Indeed, if f is strictly concave down and $x_1<x_2$, then

$$\lambda f(x_1)+(1-\lambda)f(x_2)<f(\lambda x_1+(1-\lambda)x_2)$$

for any $\lambda\in(0,1)$.

Let $a<b<c$; then $b=\lambda a+(1-\lambda)c$, where $\lambda=\frac{c-b}{c-a}\in(0,1)$. Moreover, $a-b+c=(1-\lambda)a+\lambda c$. Therefore, we have

$$\lambda f(a)+(1-\lambda)f(c)<f(b)$$

and

$$(1-\lambda)f(a)+\lambda f(c)<f(a-b+c).$$

Adding up yields

$$f(a)+f(c)<f(b)+f(a-b+c).$$

This inequality can also be viewed geometrically, noticing that the segments $[a,c]$ and $[b,a-b+c]$ (or $[a-b+c,b]$) have the same midpoint (see Figure 15.1).

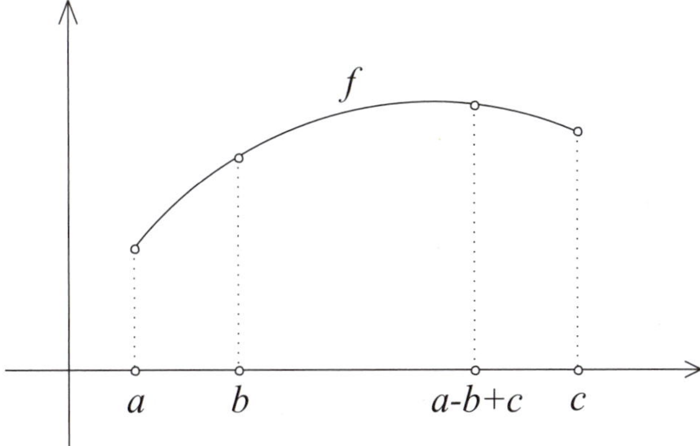

Figure 15.1

15.3. In this solution, we denote by $[\mathcal{F}]$ the set of points lying inside or on the border of figure $\mathcal{F}$. We use the following result.
Lemma. Let $KLMN$ be a convex quadrilateral. Then the set $[KLMN] - [MN]$ can be represented as a disjoint reunion of closed segments.
Proof. Suppose that KL and MN intersect at P. Then $[KLMN]-[MN]$ is the reunion of all closed segments $[XY]$, where $X \in [KN]-\{N\}$, $Y \in [LM]-\{M\}$ and P, X, Y are collinear (see Figure 15.2). If KL is parallel to MN, take XY parallel to both.

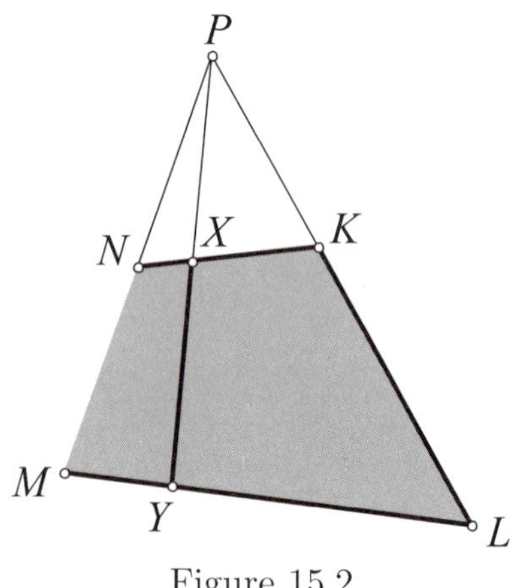

Figure 15.2

Let D, E, and F be the projections of the point T to the sides BC, CA, and AB, respectively. Then we have (see Figure 15.3)

$$S=[ABC]-\{T\} = ([FBDT] - [TD])\cup([DCET] - [ET])\cup([TEAF] - [FT]).$$

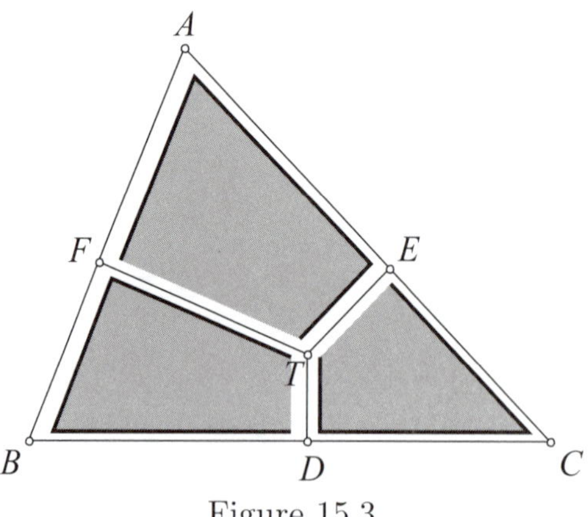

Figure 15.3

Furthermore, the sets $[FBDT] - [TD]$, $[DCET] - [ET]$, and $[TEAF] - [FT]$ are disjoint and each of them, according to the lemma, can be represented as a disjoint reunion of closed segments.

Observation. Figures 15.4 to 15.6 show how S is "filled up" with the desired closed segments.

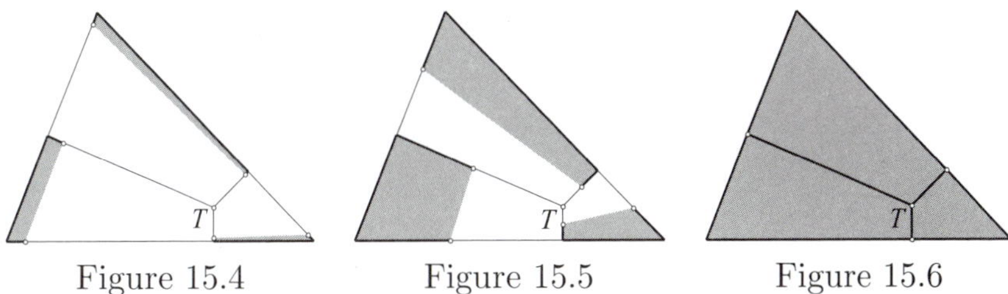

Figure 15.4 Figure 15.5 Figure 15.6

Second solution. The line segments TA, TB, and TC divide the triangle into three smaller triangles and we can see that

$$S = [ABC] - \{T\} = ([ABT] - [AT]) \cup ([ACT] - [CT]) \cup ([BCT] - [BT])$$

(see Figure 15.7). Moreover, the sets $[ABT] - [AT]$, $[ACT] - [CT]$, and $[BCT] - [BT]$ are clearly disjoint.

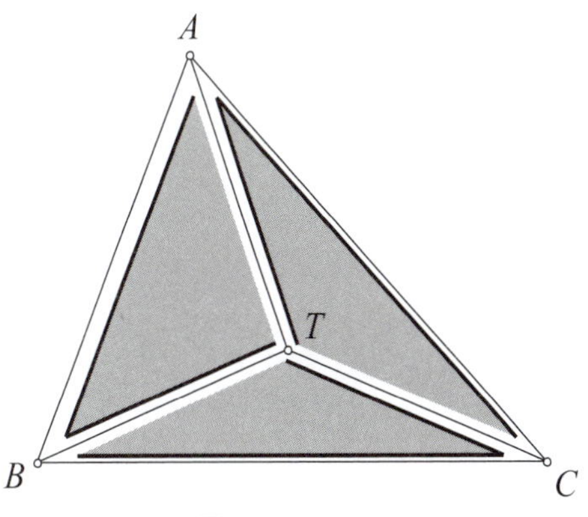

Figure 15.7

Thus, we have to show how a triangle without a side can be "filled up" with closed segments. Let KLM be the triangle and let K' and M' be the midpoints of the sides LM and KL. Consider a variable point V on the line segment KM, not equal to M (see Figure 15.8). Suppose that LV and $K'M'$ intersect at V' and let W be a point on the line segment LM such that $V'W$ is parallel to LM (if $V = K$, take $W = M'$). Since $V \neq M$, it follows that $W \neq L$.

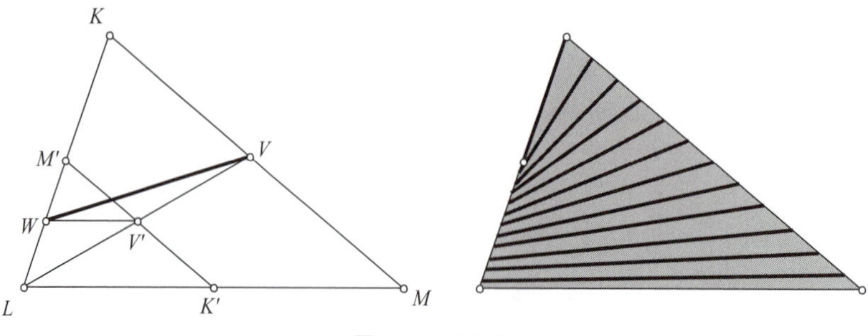

Figure 15.8

We can see that $[KLM] - [LM]$ is the reunion of all closed segments $[VW]$.

15.4. Write the equation under the form

$$y^2 + 4 = x^5.$$

The remainders of y^2 when divided by 11 are $0, 1, 3, 4, 5$, or 9. Therefore, the remainders of $y^2 + 4$ when divided by 11 are 2,4,5,7,8, or 9. Squaring both sides yields

$$\left(y^2 + 4\right)^2 = x^{10}.$$

The remainders of $\left(y^2 + 4\right)^2$ when divided by 11 are $3, 4, 5$, or 9. But, using Fermat's little theorem, we deduce that the remainders of x^{10} when divided by 11 are 0 or 1. This shows that the equation has no integer solutions.

The 16th BMO

The sixteenth BMO was held in Ochrid, Macedonia, between May 5th and May 10th, 1999. Due to the Kosovo war, near the border between Yugoslavia and Macedonia, only four countries attended this competition. Nevertheless, this BMO was very successful and offered to the participants a very pleasant stay on the Ochrid lake side. The participating countries were Albania, Bulgaria, Macedonia and Romania.

Problems

16.1. Given an acute triangle ABC, let D be the midpoint of the arc BC of its circumcircle not containing point A. The points which are symmetric to D with respect to the line BC and the circumcenter are denoted by E and F, respectively. Finally, let K be the midpoint of the line segment EA. Prove that:

a) the circle which passes through the midpoints of the sides of triangle ABC also passes through K;

b) the line which passes through K and the midpoint of BC is perpendicular to AF.

(Turkey)

16.2. Let p be a prime number such that 3 divides $p - 2$. Let

$$S = \left\{y^2 - x^3 - 1 \mid x, y \text{ are integers and } 0 \leq x, y \leq p - 1\right\}.$$

Prove that at most p elements of the set S are divisible by p.

(Bulgaria)

16.3. Let ABC be an acute triangle and let M, N, P be the orthogonal projections of the centroid G on the sides AB, BC and CA, respectively. Prove that

$$\frac{4}{27} < \frac{[MNP]}{[ABC]} \leq \frac{1}{4}.$$

(Albania)

16.4. Let $0 \leq x_0 \leq x_1 \leq \ldots \leq x_n \leq \ldots$ be an increasing sequence of nonnegative integers such that for every $k \geq 0$, the number of terms of the sequence which are not greater than k is finite, say y_k. Prove that for all positive integers m, n the following inequality holds:

$$\sum_{k=0}^{n} x_i + \sum_{j=0}^{m} y_j \geq (n+1)(m+1).$$

(Romania)

Solutions

16.1. a) Observe that the circle which passes through the midpoints A', B', C' of the sides of triangle ABC is the image of the circumcircle through a homothety centered at H, of ratio $\frac{1}{2}$. Indeed, let A'' be the point on the circumcircle such that AA'' is a diameter. Then (see Figure 16.1) the triangles BCI and $AA''C$ are right triangles, and hence

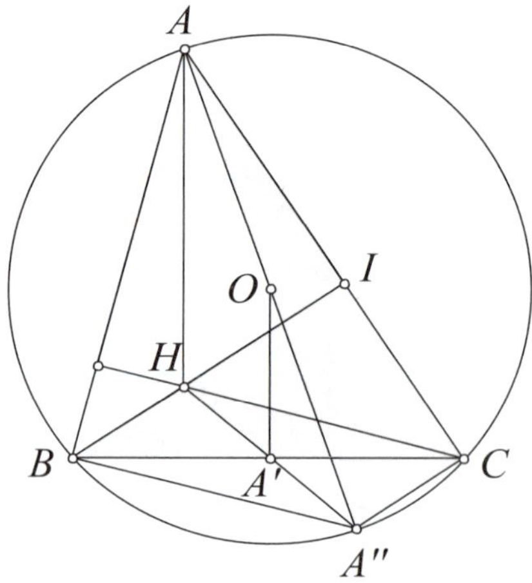

Figure 16.1

$$\angle HBC = 90^\circ - \angle ACB = \angle BCA''.$$

It follows that HB is parallel to $A''C$. Similarly, HC is parallel to $A''B$, so that the quadrilateral $A''BHC$ is a parallelogram. This implies that A' is the midpoint of the line segment HA'' and thus, A' is the image of the point A'' through a homothety of ratio $\frac{1}{2}$, centered at H. Clearly, the same homothety sends the circumcircle of triangle ABC to the circle passing to the midpoints of the triangle's sides. Also, note that in the triangle AHA'', OA' is a midline and hence, $AH = 2OA'$.

Now, since ABC is an acute triangle, O lies in its interior, and, denoting by R the circumradius of ABC, we have

$$EF = 2R - 2A'D = 2\left(R - A'D\right) = 2OA' = AH.$$

Since the line segments EF and AH are also parallel, it follows that the quadrilateral $AFEH$ is a parallelogram and K is the intersection point of its diagonals. This shows that $HK = \frac{1}{2}HF$; therefore, K belongs to the circle passing to the midpoints of the triangle's sides.

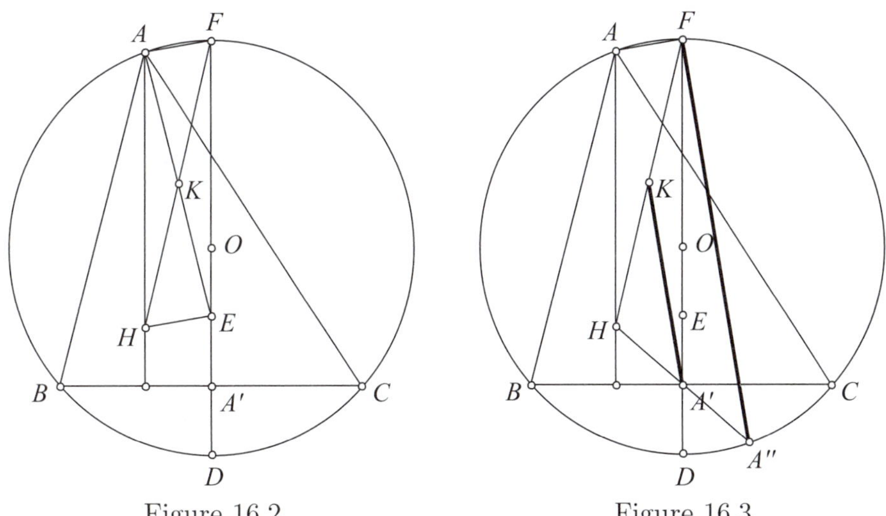

Figure 16.2 Figure 16.3

b) We saw that A' is the midpoint of the line segment HA'' and K is the midpoint of the line segment HF. Thus, in the triangle $HA''F$, $A'K$ is parallel to $A''F$. But since AA'' is a diameter, the angle $\angle AFA''$ is a right angle. It follows that $A'K$ is perpendicular to AF, as desired (see Figure 16.3).
Observation. The circle passing to the midpoints of a triangle's sides is known as the nine point circle (Euler or Feuerbach circle). For more information concerning the nine point circle, see Appendix A.

Second solution. We use complex numbers. Suppose that the circumcircle of triangle ABC is the unit circle in the complex plane and denote by $a, b, c, \ldots$ the complex affixes of the points $A, B, C, \ldots$. Then $h = a+b+c$ and $d+e = b+c$. The midpoints of the triangle's sides have affixes

$$\frac{b+c}{2}, \quad \frac{a+c}{2}, \quad \frac{a+b}{2},$$

respectively. Also, we have

$$k=\frac{a+e}{2}=\frac{a}{2}+\frac{b+c-d}{2}=\frac{a+b+c-d}{2}.$$

Let Ω be the center of the nine point circle. It is clear that

$$\omega=\frac{a+b+c}{2},$$

since

$$\left|\omega-\frac{a+b}{2}\right|=\left|\omega-\frac{b+c}{2}\right|=\left|\omega-\frac{c+a}{2}\right|=\frac{|a|}{2}=\frac{1}{2}.$$

Moreover,

$$|\omega-k|=\left|\frac{a+b+c}{2}-\frac{a+b+c-d}{2}\right|=\frac{|d|}{2}=\frac{1}{2},$$

since D lies on the circumcircle. It follows that K belongs to the circle passing through the midpoints of the sides.

For the second part of the problem, we have to prove that the real part of the number

$$\frac{a-f}{k-a'}=\frac{a+d}{\frac{a-d}{2}}$$

equals zero. This is equivalent to

$$\frac{\overline{a}+\overline{d}}{\overline{a}-\overline{d}}=-\frac{a+d}{a-d},$$

or, after a short computation,

$$|a|^2=|d|^2,$$

an obvious equality.

This solution shows that it is not necessary that ABC is an acute triangle.

16.2. We work modulo p. Let $\mathbb{F}_p$ denote the finite field of classes of integers modulo p. We use the following well known result:
Lemma. The field $\mathbb{F}_p$ contains $\frac{p-1}{2}$ non zero perfect squares.
Proof. Consider the function $f:\mathbb{F}_p\to\mathbb{F}_p$, $f(x)=x^2$. Then x and $-x$ give the same element. Moreover, if $x^2=y^2$, then $(x-y)(x+y)=0$, so either $x=y$, or $x=-y$. Thus, the image of f contains half of the elements of $\mathbb{F}_p^*$, that is, $\frac{p-1}{2}$.

Returning to our problem, consider the function $g:\mathbb{F}_p\to\mathbb{F}_p$, $g(x)=x^3+1$. This function is one to one, therefore bijective. Indeed, if $x,y\neq 0$ and $x^3+1=y^3+1$, then $x^3=y^3$ and hence $\left(xy^{-1}\right)^3=1$. Since $(p-1,3)=1$, the group $\mathbb{F}_p^*$ does not contain elements of order 3, so $xy^{-1} = 1$, that is, $x = y$. When $x^3 + 1 = 0^3 + 1 = 1$, one obtains $x = 0$.

When x runs over all elements of F_p, $x^3 + 1$ runs over all elements of F_p, as well. Among them, there are $\frac{p-1}{2}$ which are nonzero perfect squares; each element of this form is the square of precisely two elements. So, an equality of the form $y^2 = x^3 + 1$ can hold for $p - 1$ nonzero elements of F_p. Adding the case $0^2 = 0$ gives the result.

16.3. Let A' be the midpoint of the line segment BC and let K be the orthogonal projection of the point A on BC (see Figure 16.4).

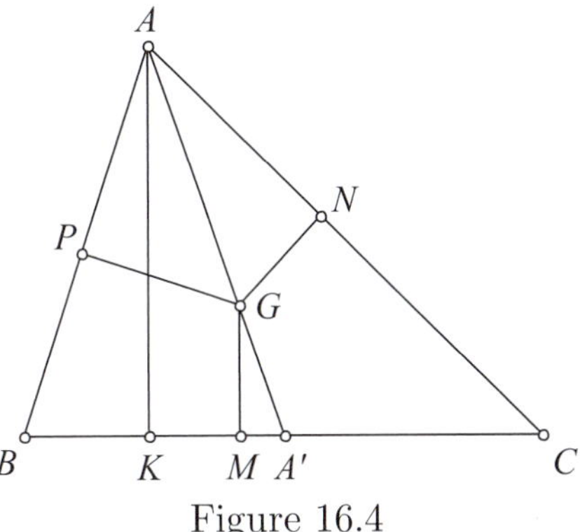

Figure 16.4

In the triangle AKA', GM is parallel to AK and we have $GM = \frac{1}{3}AK = \frac{1}{3}h_a$. Similarly, $GN = \frac{1}{3}h_b$ and $GP = \frac{1}{3}h_c$. Let S denote the area of ABC. Using the sine law and cyclic summations, we have

$$\begin{aligned}[MNP] &= \sum[MGN] = \frac{1}{18}\sum h_a h_b \sin C \\ &= \frac{1}{18}\sum \frac{4S^2}{ab}\sin C = \frac{1}{18}\sum \frac{4S^2 c}{abc}\sin C \\ &= \frac{1}{18}\sum \frac{4S^2}{4RS}\cdot\frac{c^2}{2R} = \frac{1}{36}S\sum\frac{c^2}{R}.\end{aligned}$$

Thus, we have to prove the inequalities

$$\frac{4}{27} < \frac{a^2+b^2+c^2}{36R^2} \le \frac{1}{4}.$$

The upper bound is equivalent to

$$a^2 + b^2 + c^2 \le 9R^2,$$

a consequence of the well known equality

$$OH^2 = 9R^2 - \left(a^2 + b^2 + c^2\right),$$

which holds in any triangle (see Appendix, "Leibniz Relation"). It is worth mentioning that for this inequality, the hypothesis that ABC is acute angled is not required.

For the lower bound, the following argument works: since ABC is an acute triangle, the orthocenter H lies in its interior, hence $0 \leq OH < R$.

Then the inequality $OH^2 = 9R^2 - \left(a^2 + b^2 + c^2\right) < R^2$ yields

$$\frac{4}{27} < \frac{2}{9} < \frac{a^2 + b^2 + c^2}{36R^2}.$$

Second solution. Using the sine law, the double inequality

$$\frac{2}{9} < \frac{a^2 + b^2 + c^2}{36R^2} \leq \frac{1}{4}$$

becomes

$$2 < \sin^2 A + \sin^2 B + \sin^2 C \leq \frac{9}{4}. \tag{1}$$

Using a few basic trigonometric formulae, we obtain

$$\begin{aligned}
\sin^2 A + \sin^2 B + \sin^2 C &= \frac{1 - \cos 2A}{2} + \frac{1 - \cos 2B}{2} + 1 - \cos^2 C \\
&= 2 - \frac{\cos 2A + \cos 2B}{2} - \cos^2 C \\
&= 2 - \cos(A + B)\cos(A - B) - \cos^2 C \\
&= 2 + \cos C \cos(A - B) - \cos^2 C.
\end{aligned}$$

The lower bound in (1) is now easy to prove:

$$2 < 2 + \cos C \cos(A - B) - \cos^2 C$$

is equivalent to

$$0 < \cos C \left(\cos\left(A - B\right) - \cos C\right)$$

or

$$0 < 2 \cos A \cos B \cos C,$$

which holds in any acute triangle.

For the upper bound, observe that

$$2 + \cos C \cos(A - B) - \cos^2 C \leq \frac{9}{4}$$

can be written as

$$0 \leq 4\cos^2 C - 4\cos C\cos(A - B) + 1,$$

which is equivalent to the obvious

$$0 \leq (2\cos C - \cos(A - B))^2 + \sin^2(A - B).$$

Third solution. We prove the double inequality

$$8 < \frac{a^2 + b^2 + c^2}{R^2} \leq 9.$$

Suppose that $a \geq b \geq c$ and let x be the distance from O to the side BC (see Figure 16.5). Clearly, the triangle's largest side cannot be shorter than the side of an equilateral triangle inscribed in the same circle; therefore $x \leq \frac{R}{2}$. Now, assume that the points B and C are fixed and consider the points A_1 and A_2 on the large arc BC such that $CA_1 = BA_2 = BC$. Because $AB \leq BC$ and $AC \leq BC$, the point A lies on the small arc A_1A_2. The segment AD is a median in the triangle ABC, therefore

$$4AD^2 = 2\left(AB^2 + AC^2\right) - BC^2.$$

We deduce that $b^2 + c^2$ reaches its minimal value in the same time with AD, that is, when A coincides with either A_1 or A_2 (observe that the points on the small arc A_1A_2 are outside the circle with center D and radius A_1D). Suppose $A = A_1$; then $b = a$ and $c = \frac{2ax}{R}$. Using the equality $a = 2\sqrt{R^2 - x^2}$, we obtain

$$a^2 + b^2 + c^2 = 2a^2 + \frac{4a^2x^2}{R^2} = 8R^2\left(1 - \frac{x^2}{R^2}\right)\left(1 + \frac{2x^2}{R^2}\right).$$

Denoting $t = \frac{x^2}{R^2}$, we have $0 < t \leq \frac{1}{4}$ and

$$a^2 + b^2 + c^2 = 8R^2\left(-2t^2 + t + 1\right).$$

The function $f : (0, \frac{1}{4}] \to \mathbb{R}$ is strictly increasing, therefore

$$a^2 + b^2 + c^2 > 8R^2 f(0) = 8R^2,$$

and hence

$$8 < \frac{a^2 + b^2 + c^2}{R^2}.$$

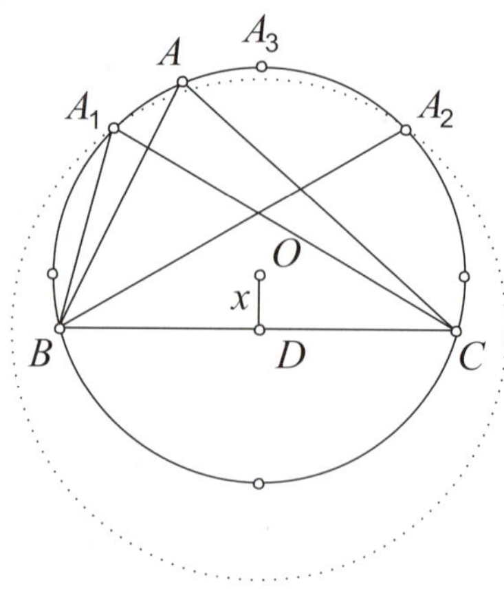

Figure 16.5

For the second inequality, observe that $b^2 + c^2$ reaches its maximal value when $A = A_3$, the midpoint of the arc A_1A_2. In this case, we obtain

$$a^2 + b^2 + c^2 = 2R^2\left(-2t^2 + 2t + 4\right),$$

where $t = \frac{x}{R}$. We have $0 < t \leq \frac{1}{2}$ and we notice that the maximal value of the function $g : (0, \frac{1}{2}] \to \mathbb{R}$, $g(t) = -2t^2 + 2t + 4$ is obtained for $t = \frac{1}{2}$ and equals $\frac{9}{2}$. Therefore

$$a^2 + b^2 + c^2 \leq 2R^2\frac{9}{2} = 9R^2,$$

which implies

$$\frac{a^2 + b^2 + c^2}{R^2} \leq 9,$$

as desired.
Observation. The inequality

$$\frac{2}{9} < \frac{a^2 + b^2 + c^2}{36R^2}$$

cannot be improved. Indeed, as a limit case, if ABC is a right triangle, with $A = 90°$, then $a^2 = b^2 + c^2 = 4R^2$ and hence

$$\frac{a^2 + b^2 + c^2}{36R^2} = \frac{8R^2}{36R^2} = \frac{2}{9}.$$

16.4. First, observe that the sequence $(y_k)_{k\geq 0}$ is also increasing and it has the "dual property": the number of terms of the sequence $(y_k)_{k\geq 0}$ which are no greater than k equals x_k. Indeed, if we have

$$0 \leq y_0 \leq y_1 \leq \ldots \leq y_{i-1} \leq k < y_i,$$

then at most k terms of the sequence $(x_n)_{n\geq 0}$ are less than or equal to i; therefore $x_k = i$.

Let m and n be positive integers. Because the number of terms of the sequence $(x_n)_{n\geq 0}$ which are equal to k is $y_k - y_{k-1}$, we have

$$x_0 + x_1 + \ldots + x_n =$$
$$= (y_1 - y_0) + 2(y_2 - y_1) + \ldots + (a-1)(y_{a-1} - y_{a-2}) + ar_a,$$

where $a = x_n$ and r_a is the number of elements from the set $\{x_0, x_1, \ldots, x_n\}$ which are equal to a. Obviously, $r_a = n + 1 - y_{a-1}$.

In the same way we obtain

$$y_0 + y_1 + \ldots + y_m =$$
$$= (x_1 - x_0) + 2(x_2 - x_1) + \ldots + (b-1)(x_{b-1} - x_{b-2}) + br_b,$$

where $b = y_m$ and $r_b = m + 1 - x_{b-1}$.

Adding up yields

$$2(x_0 + x_1 + \ldots + x_n + y_0 + y_1 + \ldots + y_m) =$$
$$= a(n+1) + b(m+1) + x_b + x_{b+1} + \ldots + x_n + y_a + y_{a+1} + \ldots + y_m.$$

For $k \geq b$, $x_k \geq x_b \geq m+1$, and because $y_m = b$, in the same way, $y_k \geq n+1$, for $k \geq a$.

It follows that

$$2(x_0 + x_1 + \ldots + x_n + y_0 + y_1 + \ldots + y_m) \geq$$
$$\geq (n+1) + (m+1) + (n+1-b)(m+1) + (m+1-a)(n+1) =$$
$$= 2(n+1)(m+1),$$

as desired.

Second solution. We use induction on $n + m$. The base case is easy, so assume that the inequality

$$\sum_{k=0}^{s} x_i + \sum_{j=0}^{t} y_j \geq (s+1)(t+1)$$

holds for all values of s, t such that $s + t \leq m + n$. We have to prove both inequalities

$$\sum_{k=0}^{n+1} x_i + \sum_{j=0}^{m} y_j \geq (n+2)(m+1) \tag{2}$$

and

$$\sum_{k=0}^{n} x_i + \sum_{j=0}^{m+1} y_j \geq (n+1)(m+2). \qquad (3)$$

For the first one, observe that if $x_{n+1} \geq m+1$, then (2) simply follows from

$$\sum_{k=0}^{n} x_i + \sum_{j=0}^{m} y_j \geq (n+1)(m+1). \qquad (4)$$

Suppose that $x_{n+1} = m-k$, with $0 \leq k \leq m$. Then $y_{m-k}, y_{m-k+1}, \ldots, y_m \geq n+2$ (at least $n+2$ terms of the sequence (x_n) are less than or equal to $m-k$). It follows that

$$y_{m-k} + y_{m-k+1} + \ldots + y_m \geq (k+1)(n+2).$$

By the induction hypothesis, we have

$$\sum_{k=0}^{n+1} x_i + \sum_{j=0}^{m-k-1} y_j \geq (n+2)(m-k).$$

Adding up the last two inequalities yields

$$\sum_{k=0}^{n+1} x_i + \sum_{j=0}^{m} y_j \geq (n+2)(m-k) + (k+1)(n+2) = (n+2)(k+1),$$

as needed.

For (3), observe that if $y_{m+1} \geq n+1$, the desired inequality also follows from (4). Suppose that $y_{m+1} = n-k$, with $0 \leq k \leq n$.

Then $x_{n-k}, x_{n-k+1}, \ldots, x_n \geq m+2$ (only the first $n-k$ terms of the sequence (x_n) are less than or equal to $m+1$) and hence

$$x_{n-k} + x_{n-k+1} + \ldots + x_n \geq (m+2)(k+1).$$

From the induction hypothesis we have

$$\sum_{k=0}^{n-k-1} x_i + \sum_{j=0}^{m+1} y_j \geq (n-k)(m+2),$$

and adding up yields

$$\sum_{k=0}^{n} x_i + \sum_{j=0}^{m+1} y_j \geq (n-k)(m+2) + (m+2)(k+1) = (n+1)(m+2).$$

Observation. A geometric interpretation of the two sequences shows that the stated inequality is a discrete version of Young's inequality for integrals (see Appendix).

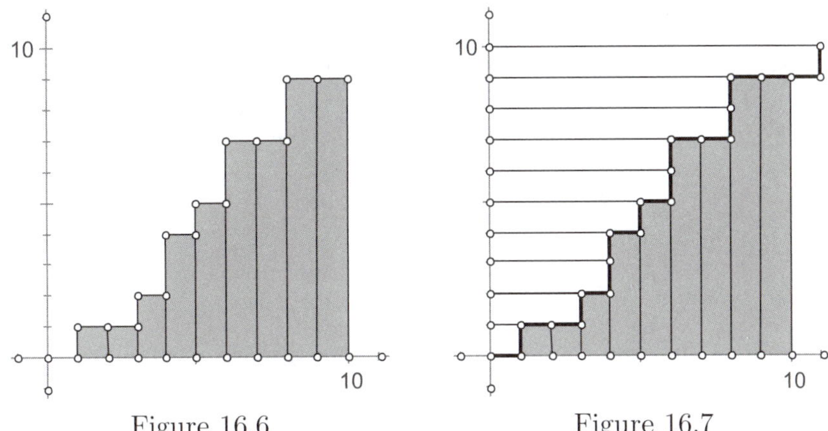

Figure 16.6 Figure 16.7

Indeed, consider x_k as the area of a rectangle erected on the Ox axis, its base being the interval $[k, k+1]$ and its height being equal to x_k (see Figure 16.6; the first terms are in this case $0, 1, 1, 2, 4, 5, 7, 7, 9, 9$). It is not difficult to see that the terms of (y_n) equal the areas of the horizontal rectangles that fit between the Oy axis and the $x-$rectangles (in Figure 16.7, we see that the first terms of (y_n) are, indeed, $1, 3, 4, 4, 5, 6, 6, 8, 8, 11, \ldots$). The inequality follows from obvious reasons (see Figure 16.8, in which $n = 8$ and $m = 4$).

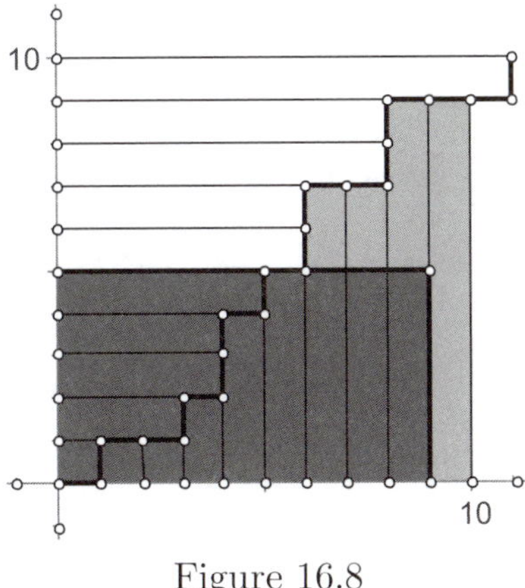

Figure 16.8

Also, we can see that the equality is reached when the point (n, m) lies on the border between the $x-$region and the $y-$region.

The 17th BMO

The seventeenth Balkan Mathematical Olympiad for high-school students was held between May 3rd and May 9th , 2000, in Chişinău, Republic of Moldova. The number of the participating countries was 9: Albania, Bulgaria, Cyprus, Former Yugoslav Republic of Macedonia, Greece, Republic of Moldova, Romania, Turkey and Yugoslavia.

Problems

17.1. Find all functions $f : \mathbb{R} \to \mathbb{R}$ satisfying the property

$$f\left(xf\left(x\right)+f\left(y\right)\right)=\left(f\left(x\right)\right)^{2}+y,$$

for all real numbers x and y.

(Albania)

17.2. Let ABC be a nonisosceles acute triangle, let D be the midpoint of the line segment BC and let E be an interior point of the median AD. The point F is the orthogonal projection of E on the line BC. Let M be an interior point of the line segment EF, and let N, P be the orthogonal projections of the point M on the lines AC and AB, respectively. Prove that the angle bisectors of $\angle PMN$ and $\angle PEN$ are parallel.

(F.Y.R. of Macedonia)

17.3. Find the maximum number of $1 \times 10\sqrt{2}$ rectangles which can be cut off from a 50×90 rectangle, by using cuts which are parallel to the edges of the given rectangle.

(Yugoslavia)

17.4. We call a positive integer r a *power* if it has the form $r = t^s$, where t, s are integers, $t, s \geq 2$.

Show that for any integer n there exists a set A of positive integers which satisfies the conditions:

(i) A has n elements;

(ii) each element of A is a power;

(iii) for any $r_1, r_2, \ldots, r_k$ $(2 \leq k \leq n)$ from A, the number

$$\frac{r_1 + r_2 + \ldots + r_k}{k}$$

is a power.

(Romania)

Solutions

17.1. This problem is identical with problem 14.4. This situation may be considered to be a weak point of the BMOs[4].

17.2. A parallel to BC through the point E intersects the sides AB and AC at B' and C', respectively. It is not difficult to see that E is the midpoint of the line segment $B'C'$ and hence ME is the perpendicular bisector of $B'C$. It follows that $MB'C'$ is an isosceles triangle, so $\angle MB'E = \angle MC'E$.

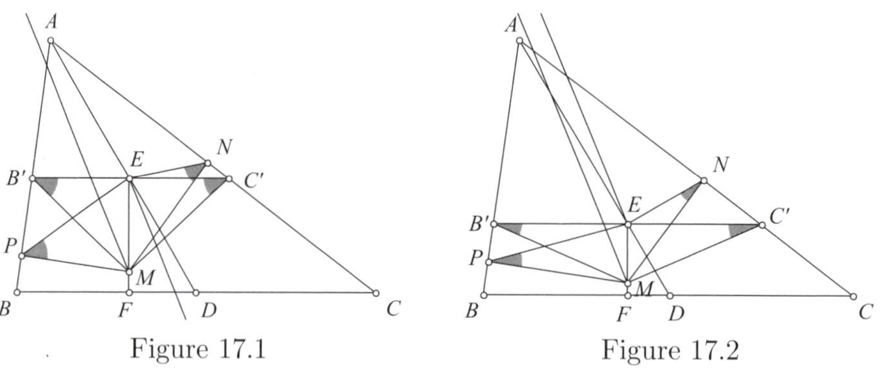

Figure 17.1

Figure 17.2

Since $\angle MPB' = \angle MEB' = 90°$, the quadrilateral $MPB'E$ is cyclic; thus, $\angle MPE = \angle MB'E$. Similarly, $\angle MNE = \angle MC'E$. We derive that in the quadrilateral $MPEN$ the opposite angles $\angle MPE$ and $\angle MNE$ are equal. This quadrilateral may be convex (as in Figure 17.1) or concave (as in Figure 17.2), but in both cases, we will prove that the angle bisectors of $\angle PMN$ and $\angle PEN$ are parallel.

If $MPEN$ is convex , suppose that the angle bisector of $\angle PMN$ intersects the line segment EP at K (see Figure 17.3). Adding up all angles of $MPEN$ yields

$$2\alpha + 2\beta + 2\gamma = 360°,$$

[4]Such unsatisfactory situation shows how useful is to have printed text or files which contain proposed problems in mathematical competitions.

hence

$$\alpha+\beta+\gamma=180^\circ.$$

This implies that $\angle PKM=\beta$ and therefore, the angle bisectors of $\angle PMN$ and $\angle PEN$ are parallel.

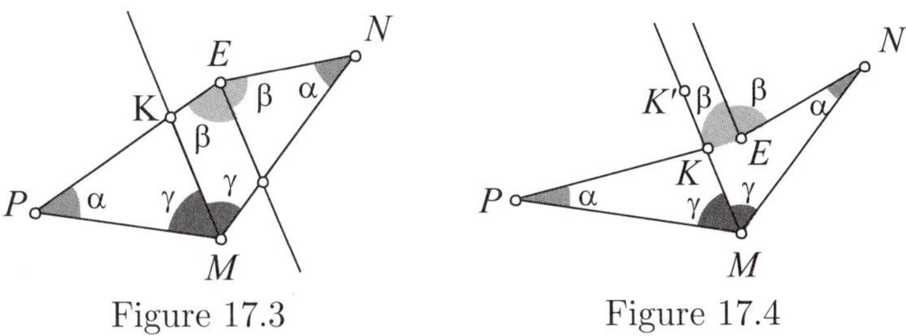

Figure 17.3 Figure 17.4

If $MPEN$ is concave (see Figure 17.4) then we have

$$2\alpha+(360^\circ-2\beta)+2\gamma=360^\circ,$$

therefore

$$\alpha+\gamma=\beta.$$

This means that the external angle $\angle PKK'$ equals β and the conclusion follows easily.

17.3. We may assume that the rectangle $ABCD$ has its vertices at the lattice points $A(0,0)$, $B(0,50)$, $C(90,50)$ and $D\,(90,0)$. In this rectangle we have 50 strips of dimensions 1×90 which are parallel to AD and BC. Because

$$60\sqrt{2}<90<70\sqrt{2},$$

each such strip contains 6 horizontal $1\times 10\sqrt{2}$ rectangles. Thus, considering the points $E(60\sqrt{2},0)$ and $F\left(60\sqrt{2},50\right)$, the rectangle $AEFB$ is covered by $6\times 50=300$ rectangles of dimensions $1\times 10\sqrt{2}$ (see Figure 17.5).

It remains to cover with $1\times 10\sqrt{2}$ strips the rectangle $EFCD$. Since

$$5<90-60\sqrt{2}<6$$

and

$$30\sqrt{2}<50<40\sqrt{2},$$

we can cover it with 5 vertical $1\times 30\sqrt{2}$ strips and then cut each strip into 3 smaller $1\times 10\sqrt{2}$ rectangles. This means that we can cut off from the original rectangle a total of 300+15=315 rectangles of dimensions $1\times 10\sqrt{2}$.

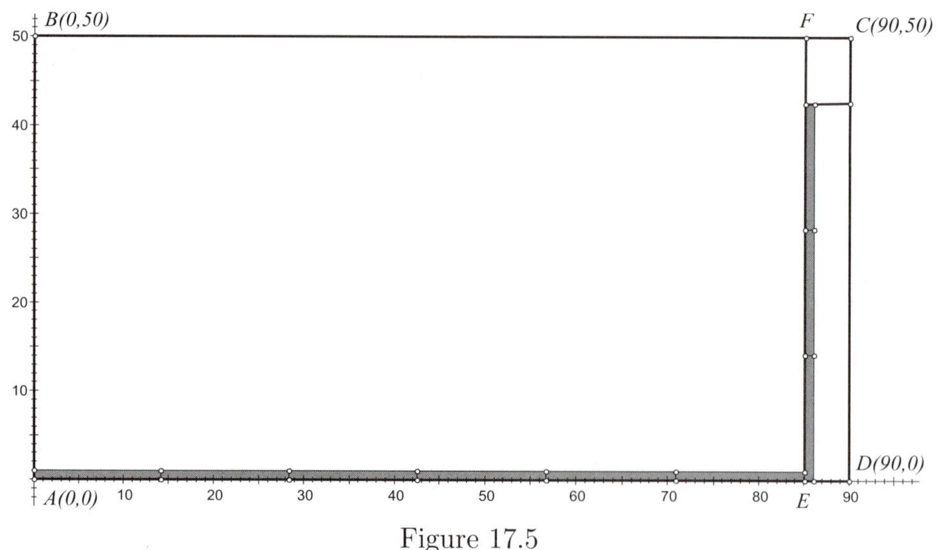

Figure 17.5

We will prove that this number 315 is maximal. Let us consider the family of lines

$$x + y = 10k\sqrt{2},$$

where $k \geq 1$ is an integer, as long as they intersect the rectangle $ABCD$ (see Figure17.6).

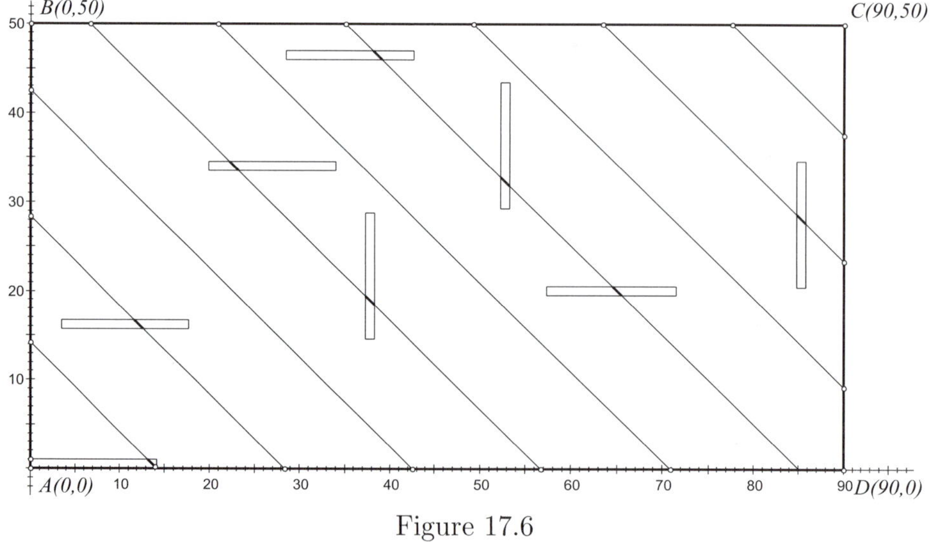

Figure 17.6

The total length of the segments determined by these lines inside the rectangle $ABCD$ is $570\sqrt{2} - 360$. When these segments intersects a $1 \times 10\sqrt{2}$ rectangle with sides parallel to the sides of $ABCD$, then the total length of the segments determined inside the rectangle equals $\sqrt{2}$. This is obvious if a

$1 \times 10\sqrt{2}$ rectangle is intersected by only one segment, but it is easy to see that the same holds if the small rectangle is intersected by two segments (see Figure 17.7).

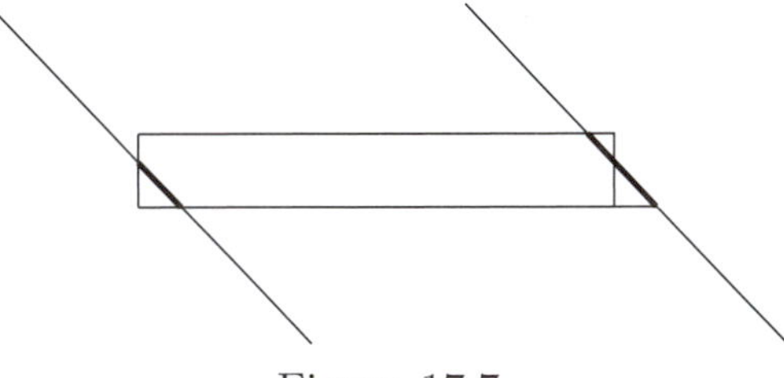

Figure 17.7

It is clear that the number of rectangles is not greater than

$$\left\lfloor \frac{570\sqrt{2} - 360}{\sqrt{2}} \right\rfloor = \left\lfloor 570 - 180\sqrt{2} \right\rfloor = 315.$$

Therefore, the family of rectangles constructed above is maximal and the required number is 315.

Second solution. There is an alternative approach for the second part of the proof. Draw horizontal and vertical lines $5\sqrt{2}$ units apart, dividing thus the 50×90 rectangle into 84 squares and 20 smaller rectangles. Color them black or white in the chessboard manner (see Figure 17.8).

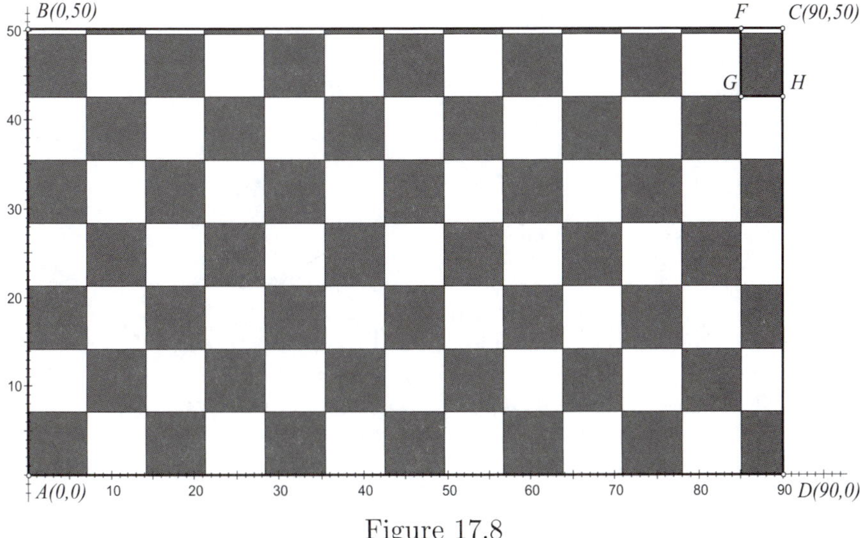

Figure 17.8

It is not difficult to see that a $1\times10\sqrt{2}$ rectangle with sides parallel to the axis covers a white area equal to $5\sqrt{2}$, regardless of its position inside the 50×90 rectangle. We want to determine the total white area. For this, observe that the surface of $ABCD$ not containing the rectangle $FCHG$ is equally divided between black and white regions. Thus, the white area outside $FCHG$ equals

$$\frac{50 \times 90 - [FCHG]}{2} = \frac{4500 - (90 - 60\sqrt{2})(50 - 30\sqrt{2})}{2} = 2850\sqrt{2} - 1800.$$

Since the area of the small white rectangle in the upright corner is

$$\left(50 - 35\sqrt{2}\right)\left(90 - 60\sqrt{2}\right) = 8700 - 6150\sqrt{2},$$

the total white area equals $6900 - 3300\sqrt{2}$. Finally, we obtain

$$\frac{6900 - 3300\sqrt{2}}{5\sqrt{2}} = 690\sqrt{2} - 660 \simeq 315.80736... < 316,$$

which proves the claim.

17.4. First, we prove the following
Lemma. Let $n \geq 1$ be a positive integer. There exists a positive integer d such that the numbers $d, 2d, \ldots, nd$ are powers.
Proof. We proceed by induction on n. For $n = 1$ or $n = 2$, take $d = 2^2$. Then 2^2, 2^3 are powers.

Suppose that the statement is true for some n. That is, there exists d such that

$$d = t_1^{s_1}, 2d = t_2^{s_2}, \ldots, nd = t_n^{s_n},$$

where $t_i, s_i \geq 2$. Let $m = \operatorname{lcm}(s_1, s_2, \ldots, s_n)$ and take $D = (n+1)^m d^{m+1}$. Then, for any i, $1 \leq i \leq n$, we have

$$iD = id\left[(n+1)d\right]^m = t_i^{s_i}\left[(n+1)d\right]^m = \left[t_i[(n+1)d]^{\frac{m}{s_i}}\right]^{s_i}$$

and $(n+1)D = [(n+1)d]^{m+1}$. Thus, the numbers

$$D, 2D, \ldots, nD, (n+1)D$$

are all powers and our lemma is proved.

Returning to the problem, we apply the lemma to the number $n \cdot n!$. There exists a number d such that the numbers d, $2d, \ldots,$ $n \cdot n!d$ are all powers. Consider the set

$$A = \{n!d,\ 2n!d, \ldots,\ n \cdot n!d\}$$

which has n elements, all of them being powers. Let $r_1, r_2, \ldots, r_k$, with $1 \leq k \leq n$, be elements of A. Then we can write $r_i = a_i n! d$, for some integers a_i, $1 \leq a_i \leq n$. We have

$$\frac{r_1 + r_2 + \ldots + r_k}{k} = \frac{n!}{k}\left(a_1 + a_2 + \ldots + a_k\right) d = md,$$

where

$$m = \frac{n!}{k}\left(a_1 + a_2 + \ldots + a_k\right)$$

is an integer. Since $a_1 + a_2 + \ldots + a_k \leq nk$, it follows that $m \leq n \cdot n!$ and hence the number

$$\frac{r_1 + r_2 + \ldots + r_k}{k}$$

is a power. Thus, the set A satisfies all required conditions.

The 18th BMO

The eighteenth Balkan Mathematical Olympiad for high-school students was held between May 3rd and May 9th, 2001, in Belgrade, Yugoslavia. The number of the participating countries was 7: Bulgaria, Cyprus, Former Yugoslav Republic of Macedonia, Greece, Republic of Moldova, Romania and Yugoslavia.

Problems

18.1. Let n be a positive integer. Show that if a and b are integers greater than 1 such that $2^n - 1 = ab$, then the number $ab - (a - b) - 1$ is of the form $k2^{2m}$, where k is odd and m is a positive integer.

(Cyprus)

18.2. We are given a convex pentagon which satisfies the following conditions:
(1) all its interior angles are congruent; and
(2) the lengths of all its sides are rational numbers.
Prove that the pentagon is regular.

(Republic of Moldova)

18.3. Let a, b, c be positive real numbers such that $a + b + c \geq abc$.
Prove that

$$a^2 + b^2 + c^2 \geq abc\sqrt{3}.$$

(Romania)

18.4. A $3 \times 3 \times 3$ cube is divided into 27 congruent unit cubic cells. One of these cells is empty and the others are filled with unit cubes, labelled in an arbitrary manner with the numbers $1, 2, \ldots 26$. An *admissible move* consists in moving an unit cube into an adjacent empty cell.

Is there a finite sequence of admissible moves after which the unit cube labelled with k and the unit cube labelled with $27 - k$ are interchanged, for each $k = 1, 2, \ldots, 13$? (Two cells are adjacent if they share a common face).

(Bulgaria)

Solutions

18.1. The problem requires to prove that the exponent of the prime 2 in the prime factorization of the number $N = ab - (a-b) - 1$ is a positive even number. The number N can be written as $N = (a+1)(b-1)$. Since a and b are odd numbers, the exponent of 2 in N is the same as the exponent of 2 in aN. We have

$$aN = (a+1)(ab-a) = (a+1)(2^n - (a+1)).$$

Since $a + 1 < 2^n$, the exponent of 2 in $a+1$ is less than n, and hence the exponent of 2 in $2^n - (a+1)$ equals the exponent of 2 in $a+1$. This proves the result.

18.2. Let $ABCDE$ be the given convex pentagon. Suppose the lines AB and DE intersect at K and the lines AB and CD intersect at L (see Figure 18.1).

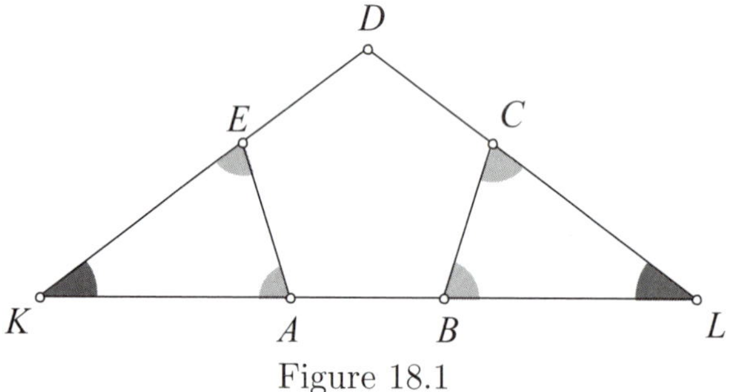

Figure 18.1

The triangles AEK and BCL are isosceles and similar, because each has two angles of $72°$ and one angle of $36°$. A trigonometric evaluation shows that

$$\frac{AE}{EK} = \frac{BC}{CL} = 2\cos 72°. \tag{1}$$

The triangle DKL is also isosceles, since $\angle AKD = \angle BLD = 36°$. Therefore, $DE + EK = DC + CL$. Suppose that $DE \neq DC$; then $EK \neq CL$ and from (1) we derive

$$2\cos 72° = \frac{AE - BC}{EK - CL} = \frac{AE - BC}{DC - DE}.$$

Since the pentagon sides have rational lengths, the above equality implies that $\cos 72°$ is a rational number. We will prove that this is not true and hence $DE = DC$. Similarly, it follows that any consecutive sides of the pentagon are equal, therefore the pentagon is regular.

Let TXY be an isosceles triangle with $\angle X = \angle Y = 72°$ and let XZ be the interior angle bisector of $\angle TXY$ (see Figure 18.2).

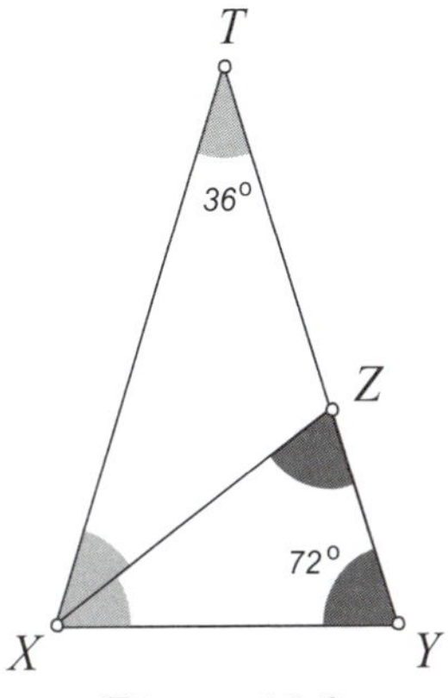

Figure 18.2

It follows that $\angle ZXY = \angle ZXT = \angle XTZ = 36°$ and $\angle XZY = \angle XYZ = 72°$. We have $XY = XZ = ZT$. Moreover, triangles TXY and XYZ are similar, hence

$$\frac{XY}{TY} = \frac{ZY}{XY}.$$

Since $ZY = TY - XY$, the above equality gives

$$\frac{XY}{TY} = \frac{TY}{XY} - 1.$$

Solving for $\dfrac{XY}{TY}$ yields

$$\frac{XY}{TY} = \frac{\sqrt{5}-1}{2}.$$

But

$$\cos 72° = \frac{1}{2}\frac{XY}{TY} = \frac{\sqrt{5}-1}{4},$$

which is an irrational number.

Second solution. We can prove a more general result: p is a prime number if and only if every polygon having equal interior angles and p sides of rational lengths is regular[5].

Suppose p is a prime number and let the rational numbers $a_1, a_2, \ldots, a_p$ be the side lengths of such a polygon. Let

$$\varepsilon = \cos\frac{2\pi}{p} + i\sin\frac{2\pi}{p}.$$

Consider the polygon's sides as vectors, oriented clockwise. Then the sum of all vectors equals zero. If we translate the vectors such that they have the

[5]This statement was published by Mihai Piticari in "Revista Matematică din Timişoara"

same origin, then the affixes of their extremities, choosing a_1 on the positive real axis, are $a_1, a_2\varepsilon, a_3\varepsilon^2, \ldots,$ and $a_n\varepsilon^{n-1}$, respectively. Since the sum of the vectors equals zero, we deduce that

$$a_1 + a_2\varepsilon + a_3\varepsilon^2 + \ldots + a_n\varepsilon^{n-1} = 0,$$

and hence ε is a root of the polynomial

$$P(X) = a_1 + a_2X + \ldots a_pX^{p-1}.$$

On the other hand, since $\varepsilon^p = 1$ and $\varepsilon \neq 1$, ε is also a root of the polynomial

$$Q(X) = 1 + X + X^2 + \ldots + X^{p-1}.$$

Since P and Q share a common root, their greatest common divisor R must be a non-constant polynomial with rational coefficients.

If the degree of R is less than $p-1$, then Q can be factorized as a product of two non-constant polynomials with rational coefficients, which is impossible (this follows from the Eisenstein criterion applied to the polynomial $Q(X+1)$ -see Appendix). It follows that $\deg R = p-1$, so there exists a real number c such that $P = cQ$. We deduce that $a_1 = a_2 = \ldots = a_p$, therefore the polygon is regular.

Conversely, suppose p is not a prime number and let $p = ab$, where a and b are positive integers greater than 1. It results that ε^a is a root of order b of the unity, and hence $1 + \varepsilon^a + \varepsilon^{2a} + \ldots + \varepsilon^{(b-1)a} = 0$. If we add this equality to $1+\varepsilon+\varepsilon^2+\varepsilon^3+\ldots+\varepsilon^{p-1} = 0$, we deduce that ε is the root of a polynomial of degree $p-1$ in which some coefficients are equal to 1 and others are equal to 2. This means that there exists a polygon with equal angles, having p sides, some of length 1 and the rest of length 2. Clearly, such a polygon is not regular.

18.3. Assume by contradiction that $a^2 + b^2 + c^2 < \sqrt{3}abc$. Then, from

$$3\sqrt[3]{a^2b^2c^2} \leq a^2 + b^2 + c^2 < \sqrt{3}abc$$

we obtain $abc > 3\sqrt{3}$. On the other hand, we have

$$\frac{a^2b^2c^2}{3} \leq \frac{(a+b+c)^2}{3} \leq a^2 + b^2 + c^2 < \sqrt{3}abc.$$

This gives $abc < 3\sqrt{3}$ an this is a contradiction.

Second solution. Using again the inequalities

$$a^2 + b^2 + c^2 \geq \frac{1}{3}(a+b+c)^2$$

and

$$a^2+b^2+c^2 \geq 3\sqrt[3]{a^2b^2c^2},$$

we obtain

$$\begin{aligned} a^2+b^2+c^2 &= \left(a^2+b^2+c^2\right)^{\frac{1}{4}}\left(a^2+b^2+c^2\right)^{\frac{3}{4}} \\ &\geq \left[\frac{1}{3}(a+b+c)^2\right]^{\frac{1}{4}}\left(3\sqrt[3]{a^2b^2c^2}\right)^{\frac{3}{4}}. \end{aligned}$$

Since $a+b+c \geq abc$, it follows that $=$

$$\begin{aligned} \left[\frac{1}{3}(a+b+c)^2\right]^{\frac{1}{4}}\left(3\sqrt[3]{a^2b^2c^2}\right)^{\frac{3}{4}} &\geq \left(\frac{a^2b^2c^2}{3}\right)^{\frac{1}{4}}\left(3\sqrt[3]{a^2b^2c^2}\right)^{\frac{3}{4}} \\ &= 3^{\frac{3}{4}-\frac{1}{4}}(abc)^{\frac{1}{2}+\frac{1}{2}} = \sqrt{3}abc, \end{aligned}$$

as desired.

Third solution. The hypothesis can be written under the form

$$\frac{1}{ab}+\frac{1}{bc}+\frac{1}{ca} \geq 1.$$

Let $\frac{1}{bc}=x$, $\frac{1}{ca}=y$, $\frac{1}{ab}=z$, where $x,y,z>0$ and $x+y+z\geq 1$.
Since $a^2=\frac{x}{yz}$, $b^2=\frac{y}{zx}$, $c^2=\frac{z}{xy}$, we have to prove that

$$x^2+y^2+z^2 \geq \sqrt{3xyz}.$$

The latter can be obtain by using $x+y+z\geq 1$ and standard inequalities:

$$\begin{aligned} x^2+y^2+z^2 &\geq \frac{(x+y+z)^2}{3} \geq \frac{(x+y+z)\sqrt{x+y+z}}{3} \\ &= \sqrt{3}\sqrt{\frac{(x+y+z)^3}{27}} \geq \sqrt{3xyz}. \end{aligned}$$

18.4. The answer is negative. Initially, we assign to each cell the number of the cube which is fitted in and to the empty cell the number 27. So, we have the assignation

$$p=\begin{pmatrix} 1 & 2 & \ldots & 26 & 27 \\ 1 & 2 & \ldots & 26 & 27 \end{pmatrix}$$

which can be interpreted as a permutation (the identity permutation). Our task is to arrive from the above p to the new permutation

$$q=\begin{pmatrix} 1 & 2 & \ldots & 26 & 27 \\ 26 & 25 & \ldots & 1 & 27 \end{pmatrix}$$

by using some transpositions of the type $(i, 27)$, where $1 \le i \le 26$.

Observe that q is an odd permutation since it decomposes into an odd number of transpositions: $q = (1,26)(2,25)\ldots(13,14)$. Therefore, if the permutation q is obtained from p, the number of transpositions of the type $(i, 27)$ must be odd.

On the other hand, if we actually look more geometrically to the admissible moves, we can see that in any such move the empty cell provides moves of the following types: up-down, left-right and forward-backward. Therefore, if at the end the empty cell is in the same place, an even number of admissible moves have been performed. This is a contradiction.

Second solution. Label the cubic cells with the numbers $1, 2, \ldots, 27$ as follows: use the numbers $1, 2, \ldots, 9$ for the lower level cells, the numbers $10, 11, \ldots, 18$ for the middle level cells and the numbers $19, 20, \ldots, 27$ for the upper level cells. The labelling is made such that we obtain the following distribution of the numbers:

7	8	9
4	5	6
1	2	3

lower level

16	17	18
13	14	15
10	11	12

middle level

25	26	27
22	23	24
19	20	21

upper level

After that we take out the empty cell and subtract one unit from the labels of the next labelled cells. For instance, if the empty cell is the cell originally labelled with 22, then we obtain the distribution:

7	8	9
4	5	6
1	2	3

lower level

16	17	18
13	14	15
10	11	12

middle level

24	25	26
	22	23
19	20	21

upper level

When we put in the cell i, $1 \le i \le 26$, the corresponding cube c_i, we obtain a permutation

$$c = \begin{pmatrix} 1 & 2 & \ldots & 26 \\ c_1 & c_2 & \ldots & c_{26} \end{pmatrix}.$$

An admissible move defines a permutation p of the set of the cubes $\{c_1, \ldots, c_{26}\}$ defined as follows: move the cube into the empty cell and then read the cube's labels in the order of the corresponding cells where they are placed. We shall see that any such permutation is even. To do this we remark that we have three types of such permutations p:

- when the admissible move is made on a line at the same level; in this case p is the identity permutation.

- when the admissible move is made on a column at the same level; in this case, p has either the form

$$p = \begin{pmatrix} \dots & a & b & c & d & \dots \\ \dots & a & d & b & c & \dots \end{pmatrix}$$

for a backwards move, or

$$p = \begin{pmatrix} \dots & a & b & c & d & \dots \\ \dots & b & c & a & d & \dots \end{pmatrix}$$

for a move ahead. In both cases, p is even.

- when the admissible move is made upwards or downwards by changing the level. For a move upwards we obtain

$$p = \begin{pmatrix} \dots & a & b & c & d & e & f & g & h & i & \dots \\ \dots & b & c & d & e & f & g & h & i & a & \dots \end{pmatrix}$$

and for a downwards move

$$p = \begin{pmatrix} \dots & a & b & c & d & e & f & g & h & i & \dots \\ \dots & i & a & b & c & d & e & f & g & h & \dots \end{pmatrix}.$$

Both permutations are 9−cycles, so they are even permutations.

The task is to arrange the cubes in the order

$$d = \begin{pmatrix} 1 & 2 & \dots & 26 \\ c_{26} & c_{25} & \dots & c_1 \end{pmatrix}$$

which is an odd permutation. This is a contradiction.

The 19th BMO

The nineteenth Balkan Mathematical Olympiad for high-school students was held between April 25th and May 1st , 2002, in Antalya, Turkey. The number of the participating countries was 9: Albania, Bulgaria, Cyprus, Former Yugoslav Republic of Macedonia, Greece, Republic of Moldova, Romania, Turkey and Serbia-Montenegro.

Problems

19.1. Let $n \geq 4$ be a positive integer and let $A_1, A_2, \ldots A_n$ be points in the plane, any three noncollinear. Some pairs of distinct points are connected by a segment in such a way that each point is incident with at least three segments.

Show that there exist distinct points X_1, X_2 , ..., X_{2k} among the points $A_1, A_2, \ldots, A_n$ such that X_i is connected to X_{i+1}, for all $i = 1, 2, \ldots, k$ (where $X_{2k+1} = X_1$).

(Yugoslavia)

19.2. Consider the sequence $(a_n)_{n \geq 1}$ defined as follows: $a_1 = 20$, $a_2 = 30$ and $a_{n+1} = 3a_n - a_{n-1}$, for all $n \geq 2$. Find all values of n for which $1 + 5a_n a_{n+1}$ is a perfect square.

(Bulgaria)

19.3. Two unequal circles intersect at the points A and B. Their common tangents touch the first circle at M, S and the second circle at N, T, respectively. Prove that the orthocenters of triangles AMN, BMN, AST, and BST are the vertices of a rectangle.

(Romania)

19.4. Find all functions $f : \mathbb{N} \to \mathbb{N}$ which satisfy the condition

$$2n + 2001 \leq f\left(f\left(n\right)\right) + f\left(n\right) \leq 2n + 2002, \forall n \in \mathbb{N}.$$

(Romania)

Solutions

19.1. The problem can be restated in the language of graph theory: given a graph $G = (V, E)$ such that every vertex has degree at least 3, show that the graph contains an even cycle.

We use induction on n, the number of vertices of the graph G. When $n = 4$, than $G = K_4$ (see Figure 19.1). In this case, 12341 is an even cycle. Let $n \geq 4$ and assume the statement is true for all graphs with less vertices.

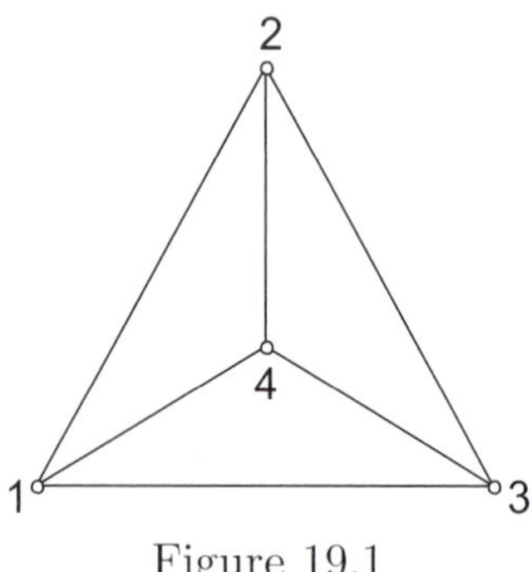

Figure 19.1

Take a vertex v_1 and construct a path $v_1v_2 \dots v_m$ along distinct edges of G. Since G is finite, we eventually find a cycle $C = v_1v_2 \dots v_sv_1$ of length s. Suppose that s is odd. Since every vertex in C has degree 2 and every vertex in G has degree at least 3, it follows that every vertex v_i of C is connected with a vertex $x \in G \backslash C$.

There are two cases:

Case 1. There exists a vertex $x \in G \backslash C$ which is connected with two distinct vertices of C, say v_1, v_i. Then one of the cycles $xv_1 \dots v_ix$ or $xv_iv_{i+1} \dots v_1x$ is even.

Case 2. Every vertex v_i of C is connected with a vertex x_i of $G \backslash C$ such that $x_1, x_2, \dots, x_s$ are distinct. Then we construct a new graph H in the following way: the vertices of H are the vertices of $G \backslash C$ and some extravertex y and the edges of H are all edges of $G \backslash C$ and all the extra edges x_iy (this means that H is obtained from G by "shrinking" C to a single vertex, y). Since $s \geq 3$ and $x_1, x_2, \dots x_s, y \in H$, it follows that H has at least 4 and less than n vertices. It is clear that every vertex of H has degree at least 3. By induction, H contains an even cycle D. If $y \notin D$, then D is an even cycle in G, as well, and we are done. If D contains y, then at least one of the cycles which contain D and a path in C is an even cycle in G.

Second solution. Let $v_1v_2\ldots v_m$ be a path of maximal length in G (recall that $v_1,\ldots,v_m$ should be distinct vertices in V). Since every vertex has the degree at least 3, there exist at least two vertices, say x and y which are connected to v_1. If either of these vertices is distinct from $v_2,v_3,\ldots,v_m$ then the path $v_1v_2\ldots v_m$ can be extended, contradicting thus its maximality. Hence there exist $i<j$ such that $x=v_i$ and $y=v_j$. Now, consider the cycles $v_1v_iv_{i+1}\ldots v_jv_1$, $v_iv_1\ldots v_{i-1}v_i$ and $v_jv_1\ldots v_{j-1}v_j$. Their lengths are $j-i+2$, i and j, respectively. The conclusion follows from the fact that i,j and $j-i+2$ cannot be all odd numbers since their sum is even.

Third solution. Let $C=v_1v_2\ldots v_mv_1$ be an odd cycle of the graph. We consider two cases:

Case 1. A vertex from C, say v_1, is connected to another vertex v_i, distinct from v_2 and v_m. We obtain the cycles $v_1v_2\ldots v_iv_1$ and $v_1v_iv_{i+1}\cdots v_mv_1$ of length i and $m-i$. Since m is odd, exactly one of these cycles is even.

Case 2. v_1 is connected to a vertex x_1, $x_1\neq v_2, x_1\neq v_m$. We construct a path of the form $v_1x_1x_2\ldots x_p$ where x_p is the first vertex which is either of the form v_1, $2\le i\le m$ or $x_p=x_j$, $j<p-1$. In both cases we apply the previous argument.

19.2. Since the sequence $(a_n)_{n\geq 1}$ is defined by a second order linear recurrence relation, it follows that

$$a_n=c_1x_1^n+c_2x_2^n,$$

where c_1,c_2 are real constants and x_1,x_2 are the roots of the characteristic equation

$$x^2-3x+1=0.$$

A straight computation shows that

$$a_n=10\left(x_1^{n-1}+x_2^{n-1}\right).$$

Then

$$\begin{aligned}5a_na_{n+1}&=500\left(x_1^{n-1}+x_2^{n-1}\right)\left(x_1^n+x_2^n\right)=\\&=500\left(x_1^{2n-1}+x_2^{2n-1}+(x_1x_2)^{n-1}(x_1+x_2)\right)=\\&=500\left(x_1^{2n-1}+x_2^{2n-1}+3\right).\end{aligned}$$

Let $y_1=\frac{1+\sqrt{5}}{2}$ and $y_2=\frac{1-\sqrt{5}}{2}$. Then y_1,y_2 are the roots of the equation $y^2-y-1=0$, which is the characteristic equation of a Fibonacci type sequence.

It is known that the classical Fibonacci sequence $(f_n)_{n\geq 1}$ is given by $f_1 = f_2 = 1$ and $f_{n+2} = f_{n+1} + f_n$, for $n \geq 1$, and that

$$f_n = \frac{1}{\sqrt{5}}\left(y_1^n - y_2^n\right).$$

Observe that $x_1 = y_1^2$ and $x_2 = y_2^2$. This implies that

$$5a_na_{n+1} = 500\left(y_1^{4n-2} + y_2^{4n-2} + 3\right) = 500\left(y_1^{4n-2} + y_2^{4n-2} + 2\right) + 500$$
$$= 10^2\left(5\left(y_1^{2n-1} - y_2^{2n-1}\right)^2\right) + 500 = 10^2\left(25\left(\frac{y_1^{2n-1} - y_2^{2n-1}}{\sqrt{5}}\right)^2\right) + 500$$
$$= 50^2 f_{2n-1}^2 + 500 = \left(50 f_{2n-1}\right)^2 + 500.$$

In our problem, we look for those n for which $5a_na_{n+1} + 1 = \left(50f_{2n-1}\right)^2 + 501$ is a perfect square. Therefore, we have to solve the Diophantine equation

$$x^2 = y^2 + 501 \iff (x-y)(x+y) = 501,$$

where x, y are positive integers. Since $501 = 1 \cdot 501 = 3 \cdot 167$, we obtain either $x = 85, y = 82$ or $x = 251, y = 250$. But y must be divisible by 5, so that the only candidate is $y = 250$ which shortly leads to $n = 3$.

Second solution. This solution is based on the quick remark which is already contained in the first solution: $5a_na_{n+1} - \left(a_n + a_{n+1}\right)^2$ is constant.

Indeed, we have

$$5a_na_{n+1} - \left(a_n + a_{n+1}\right)^2 = 5a_n\left(3a_n - a_{n-1}\right) - \left(4a_n + a_{n-1}\right)^2 =$$
$$= 5a_na_{n-1} - \left(a_n + a_{n-1}\right)^2.$$

Iterating this, we obtain

$$5a_na_{n+1} - \left(a_n + a_{n+1}\right)^2 = 5a_2a_1 - \left(a_1 + a_2\right)^2 = 500,$$

and hence

$$5a_na_{n+1} + 1 = \left(a_n + a_{n+1}\right)^2 + 501.$$

The solution ends as the previous one.

Third solution. We will basically use the same ideas. Put $b_n = a_{n+1} + a_n$ and $c_n = 1 + 5a_na_{n+1}$. Then

$$5a_{n+1} = b_{n+1} + b_n$$

and

$$a_{n+2} - a_n = b_{n+1} - b_n.$$

Hence

$$c_{n+1} - c_n = 5a_{n+1}\left(a_{n+2} - a_n\right) = b_{n+1}^2 - b_n^2.$$

Therefore, for any n we have

$$c_{n+1} - b_{n+1}^2 = c_n - b_n^2 = c_1 - b_1^2 = 501,$$

and we continue as in the first solution.

19.3. Let O_1 and O_2 be the centers of the given circles. Then the line O_1O_2 is a symmetry axis of the picture. Since the triangles AMN and BST are symmetric with respect to O_1O_2 it follows that their orthocenters H_1 and H_2 are also symmetric with respect to O_1O_2. The same is true for H_3 and H_4, the orthocenters of the triangles AST and BMN. Therefore $H_1H_2H_3H_4$ is an isosceles trapezoid or a rectangle. Then it is sufficient to prove that H_1H_3 is perpendicular to AB. We will prove more: H_1B and H_3B are perpendicular to AB.

Suppose MN intersects AB and AH_1 at Q and P, respectively. We will prove that the quadrilateral H_1BQP is cyclic. This is equivalent to

$$AH_1 \cdot AP = AB \cdot AQ.$$

The power of a point theorem applied to P and the circle AMN gives

$$PM \cdot PN = PH_1 \cdot PA,$$

and hence

$$\begin{aligned} AH_1 \cdot AP &= (AP - H_1P) \cdot AP = AP^2 - PM \cdot PN = \\ &= AQ^2 - PQ^2 - (MQ - PQ)(NQ + QP). \end{aligned}$$

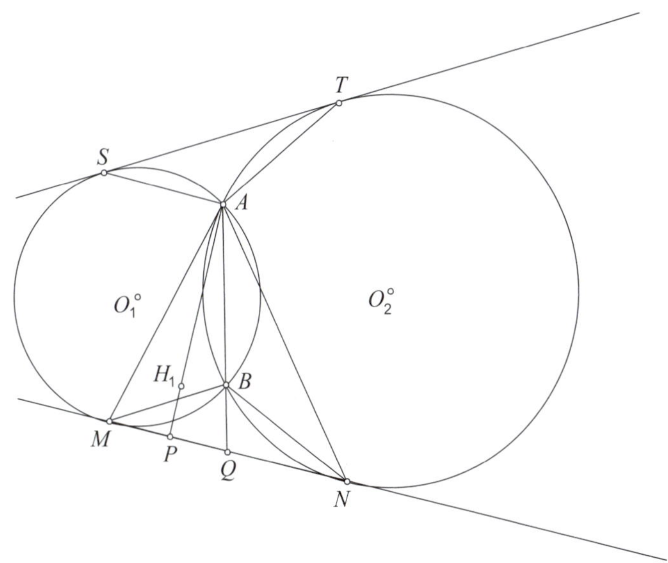

Figure 19.2

Observe that Q is the midpoint of the line segment MN. This follows from the power of a point theorem applied to Q and the two given circles: $QM^2 = QA \cdot QB = QN^2$. Therefore, we have

$$AH_1 \cdot AP = AQ^2 - PQ^2 - MQ^2 + PQ^2$$
$$= AQ^2 - AQ \cdot BQ = AQ\,(AQ - BQ) = AQ \cdot AB,$$

as desired.

19.4. We define the sequence of positive integers $(a_n)_{n\geq 0}$ as follows: a_0 is an arbitrary positive integer and $a_{n+1} = f(a_n)$, for all n. The given condition yields

$$2001 \leq a_k + a_{k-1} - 2a_{k-2} \leq 2002,$$

for all $k \geq 2$. Denote $b_k = a_k - 667k$; then

$$0 \leq b_k + b_{k-1} - 2b_{k-2} \leq 1,$$

for all $k \geq 2$. It follows that $c_k = b_k + b_{k-1} - 2b_{k-2}$ is an integer equal to either 0 or 1. Write this as

$$b_k - b_{k-1} = -2\,(b_{k-1} - b_{k-2}) + c_k$$

and iterate; we obtain

$$b_k - b_{k-1} = (-2)^{k-1}(b_1 - b_0) + (-2)^{k-2} c_2 + (-2)^{k-3} c_3 + \ldots + c_k,$$

for all $k \geq 2$. Adding up for $k = 2, 3, \ldots, n$ yields

$$b_n = \frac{2b_0 + b_1}{3} + \frac{(-2)^n}{3} + \frac{2}{3}\left(c_n - 2c_{n-1} + 2^2 c_{n-2} + \ldots + (-2)^{n-2} c_2\right) + \quad (1)$$
$$+\frac{1}{3}(c_2 + c_3 + \ldots + c_n).$$

Suppose that n is an odd integer. Then, because all c_i's are either 0 or 1, we have

$$c_n - 2c_{n-1} + \ldots + (-2)^{n-2} c_2 \leq 1 + 2 + 2^2 + \ldots + 2^{n-3} = \frac{2^{n-1} - 1}{3}$$

Let us denote $b_1 - b_0 = B$. From (1) we deduce that there exist real numbers α and β such that

$$b_n \leq \alpha + \beta n - \frac{2^n}{3}B + \frac{2}{3} \cdot \frac{2^{n-1} - 1}{3},$$

for all n. Put $\gamma = \beta + 667$ and remember that $b_n = a_n - 667n$ to obtain

$$a_n \leq \alpha + \gamma n - \frac{2^n}{3}\left(B - \frac{1}{3}\right). \tag{2}$$

Consider that $B > 0$; since B is an integer number, it follows that $B - \frac{1}{3} \geq \frac{2}{3}$. For $n > 4$ we have $2^n > n^2$ and then, using (2), we obtain

$$a_n \leq \alpha + \gamma n - \frac{4}{9}n^2,$$

for all odd $n > 4$. For large n, this contradicts the condition that a_n is a positive integer.

In the same way (taking n to be even) we prove that $B < 0$ also leads to a contradiction. We conclude that $B = 0$, $b_0 = b_1$ and $a_1 = a_0 + 667$. Since a_0 was an arbitrary integer, we obtain

$$f(n) = n + 667,$$

for all n. A straightforward computation shows that this function indeed satisfies the required condition.

Second solution. Define the sequence $(a_n)_{n\geq 0}$ as before and let

$$c_n = a_n - a_{n-1} - 667,$$

for all $n \geq 1$. Then

$$c_{n+1} + 2c_n = a_{n+1} + a_n - 2a_{n-1} - 2001,$$

therefore for all $n \geq 1$, $c_{n+1} + 2c_n$ is an integer, satisfying

$$0 \leq c_{n+1} + 2c_n \leq 1.$$

Suppose that $c_1 > 0$; then $c_1 \geq 1$ and $c_2 \leq -2c_1 + 1 \leq -1$. Moreover, $c_3 \geq -2c_2 \geq 2$. We claim that $c_{2k+1} \geq 2^k$ and we prove it by induction. If $c_{2k+1} \geq 2^k$ holds, then $c_{2k+2} \leq -2c_{k+1} + 1 \leq -2^{k+1} + 1$ and $c_{2k+3} \geq -2c_{k+2} \geq 2^{k+2} - 2 \geq 2^{k+1}$, and the claim is proved.

Observe that

$$a_{2k+2} - a_{2k} - 1334 = c_{2k+2} + c_{2k+1} \leq -2^k + 1,$$

therefore, for $k \geq 11$, we have

$$a_{2k+2} < a_{2k}.$$

This is a contradiction, since all a_k's must be positive integers.

If $c_1 < 0$, then $c_2 \geq -2c_1 > 0$ and we repeat our argument until we obtain

$$a_{2k+3} < a_{2k+1},$$

for all $k \geq 11$, another contradiction.

We conclude that $c_1 = 0$, that is, $a_1 = a_0 + 667$. As before, it follows that

$$f(n) = n + 667,$$

for all n.

The 20th BMO

The twentieth Balkan Mathematical Olympiad for high-school students was held between May 2nd and May 5th, 2003, in Tirana, Albania. The number of the participating countries was 9: Albania, Bulgaria, Cyprus, Former Yugoslav Republic of Macedonia, Greece, Republic of Moldova, Romania, Turkey and Serbia-Montenegro.

Problems

20.1. Is there a set of 4004 positive integers so that the sum of any 2003 of them is not divisible by 2003?

(F.Y.R. of Macedonia)

20.2. Let ABC be a triangle with $AB \neq BC$ and let D be the point of intersection between the tangent at A to the circumcircle of ABC and the line BC. The perpendiculars to BC erected at points B and C and the perpendicular bisectors of the sides AB and AC meet at the points E, F respectively. Prove that the points D, E, and F are collinear.

(Romania)

20.3. Find all functions $f : \mathbb{Q} \to \mathbb{R}$ satisfying the conditions:

(a) $f(1) + 1 > 0$;

(b) $f(x+y) - xf(y) - yf(x) = f(x) f(y) - x - y + xy$, for all $x, y \in \mathbb{Q}$;

(c) $f(x) = 2f(x+1) + x + 2$, for all $x \in \mathbb{Q}$.

(Cyprus)

20.4. Let $ABCD$ be a rectangle whose sides have lengths m, n, respectively. Assume that m and n are two odd relatively prime integers and that the rectangle is partitioned into $m \times n$ unit squares. The main diagonal AC intersects the sides of the unit squares at the points $A_1, A_2, \ldots, A_N$, in that order, where $N \geq 2$ and $A_1 = A$, $A_N = C$. Prove that

$$A_1A_2 - A_2A_3 + A_3A_4 - \ldots + (-1)^{N-1} A_{N-1}A_N = \frac{\sqrt{m^2+n^2}}{mn}.$$

(Bulgaria)

Solutions

20.1. The answer is yes. A possible construction is the following: take 4004 distinct numbers $a_1, a_2, \ldots, a_{2002}, b_1, b_2, \ldots, b_{2002}$ such that for all i, $1 \leq i \leq 2002$, we have $a_i \equiv 1 \pmod{2003}$ and $b_i \equiv 0 \pmod{2003}$. Then, any subset of the set $\{a_1, a_2, \ldots, a_{2002}, b_1, b_2, \ldots, b_{2002}\}$ having 2003 elements contains at least one of the a_i's and at most 2002. Therefore, the sum of its elements cannot be congruent to 0 modulo 2003.
Observation. If we replace 4004 by 4005, the answer is negative. Moreover, the following is true: from any set of $2n-1$ integers, one can pick n whose sum is divisible by n[1].

20.2. We will prove that $\angle BDE = \angle CDF$, which clearly implies that D, E, and F are collinear. To do this, it is sufficient to show that

$$\frac{BE}{CF} = \frac{BD}{CD}. \tag{1}$$

Let M be the midpoint of the line segment AB. In the triangle BME we have

$$\cos\angle MBE = \sin B = \frac{BM}{BE} = \frac{AB}{2BE},$$

hence

$$BE = \frac{AB}{2\sin B}.$$

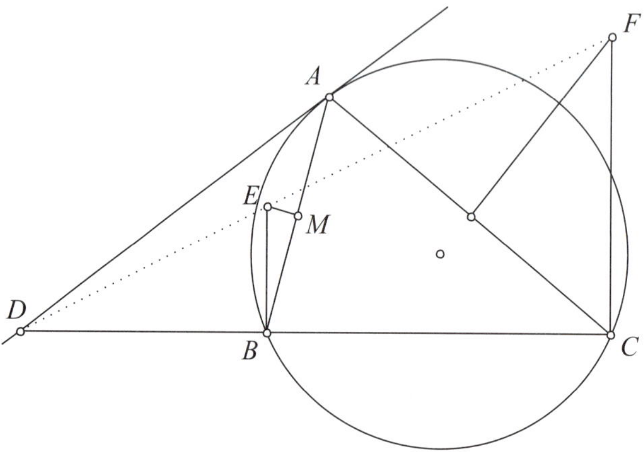

Figure 20.1

Similarly, we obtain

$$CF = \frac{AC}{2\sin C},$$

[1]See problem M45, Kvant,9,1970

therefore we derive

$$\frac{BE}{CF} = \frac{AB \sin C}{AC \sin B},$$

or, using the sine law

$$\frac{BE}{CF} = \frac{\sin^2 C}{\sin^2 B}.$$

The lengths of the line segments BD and CD can be computed using again the sine law in the triangles ABD and ADC, respectively. We have

$$\frac{BD}{\sin C} = \frac{AD}{\sin (\pi - B)},$$

$$\frac{DC}{\sin (A + C)} = \frac{AD}{\sin C}.$$

It follows that

$$\frac{BD}{CD} = \frac{\sin^2 C}{\sin^2 B},$$

proving thus (1).

Second solution. Let us consider an inversion of pole D and ratio $\rho = DA^2 = DB \cdot DC$. Then A is a fixed point of this transformation and the points B and C are interchanged. The circle $\mathcal{C}$ centered at E with radius BE is tangent to the line DC at C and passes through A. This circle is sent to a circle $\mathcal{C}'$ tangent to BC at C and passing through A as well. But then $\mathcal{C}'$ is centered at F. It follows that the point E is sent to F and hence the points D, E, and F are collinear.

20.3. The conditions (a) and (b) invite us to use the substitution $g(x) = f(x) + x$, where $g : \mathbb{Q} \to \mathbb{R}$. Thus, condition (b) becomes

$$g(x + y) = g(x) g(y), \tag{b'}$$

while from (a) we have $g(1) > 0$.

Plugging $x = y = 0$ in (b$'$) yields $g(0) = g^2(0)$, whence either $g(0) = 0$ or $g(0) = 1$. If $g(0) = 0$, then $g(1) = g(1 + 0) = g(1) g(0) = 0$, a contradiction. Therefore, we have $g(0) = 1$.

Put $g(1) = a$, with $a > 0$. Then an easy inductive argument shows that

$$g(n) = g(1)^n = a^n,$$

for all positive integers n. Also,

$$a = g(1) = g\underbrace{\left(\frac{1}{n} + \frac{1}{n} + \ldots + \frac{1}{n}\right)}_{n \text{ times}} = g^n\left(\frac{1}{n}\right)$$

whence

$$g\left(\frac{1}{n}\right) = a^{\frac{1}{n}}.$$

Moreover,

$$g\left(\frac{m}{n}\right) = g\underbrace{\left(\frac{1}{n} + \frac{1}{n} + \ldots + \frac{1}{n}\right)}_{m \text{ times}} = g^m\left(\frac{1}{n}\right) = a^{\frac{m}{n}},$$

therefore we obtain

$$g(x) = a^x,$$

for all nonnegative rational numbers x. Finally, observe that

$$1 = g(0) = g(x - x) = g(x)\, g(-x) = a^x g(-x),$$

so $g(-x) = a^{-x}$.

Consequently, we have $g(x) = a^x$, for all rational numbers x.

Condition (c) leads to

$$a^x - x = 2\left(a^{x+1} - x - 1\right) + x + 2,$$

or, equivalently,

$$a^x = 2a^{x+1}.$$

We obtain $a = \frac{1}{2}$ and it easy to check that the function

$$f(x) = \frac{1}{2^x} - x$$

indeed satisfies the required conditions.

20.4. Denote the required sum by

$$S = \sum_{i=1}^{N-1} (-1)^{i-1} A_i A_{i+1}.$$

Suppose $m > n$ and consider the given rectangle as having its vertices in the coordinate plane whose origin is $A(0,0)$ such that C has the coordinates (m, n). Also, consider the grid whose vertical lines have equations $x = k$, $1 \le k \le m-1$ and whose horizontal lines have equations $y = l$, $1 \le l \le n-1$. The diagonal AC meets the vertical lines at the points $V_k\left(k, \frac{n}{m}k\right)$, and the horizontal lines at the points $H_l\left(\frac{m}{n}l, l\right)$. Since $\gcd(m, n) = 1$, the points V_k and H_l are all distinct. Taking into account the points A and C, we obtain that the sequence $A = A_1, A_2, \ldots, A_{N-1}, A_N = C$ consists of

$$N = (m-1) + (n-1) + 2 = m + n$$

points, determining thus $m+n-1$ segments on the diagonal AC, that is, an odd number of segments.

We call a segment A_iA_{i+1} to be of the first type if it connects two points V_k and V_{k+1} (see Figure 20.2). We call a segment A_iA_{i+1} to be of the second type if it connects a point V_k and a point H_l, in some order (see Figure 20.3). The segments A_1A_2 and $A_{N-1}A_N$ will be considered to be of the first type.

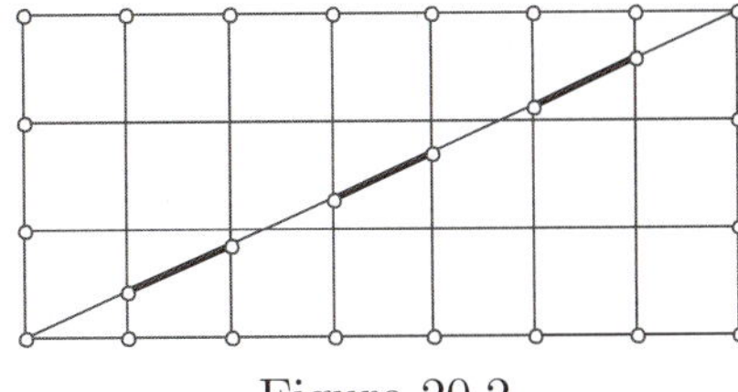
Figure 20.2

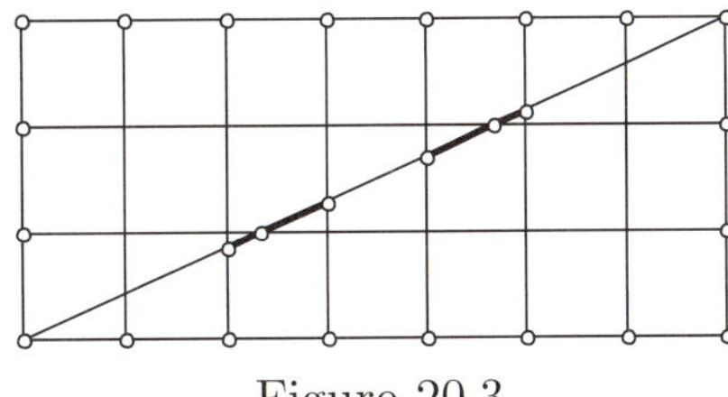
Figure 20.3

A segment of the first type has the standard length $\dfrac{\sqrt{m^2+n^2}}{m}$. A segment of the second type appears when a point H_l is lies between two points V_k and V_{k+1}.. Hence, in the sum S, one necessarily has two consecutive segments the second type and they contribute to the sum with distinct signs: either $A_iA_{i+1} - A_{i+1}A_{i+2}$ or $-A_iA_{i+1} + A_{i+1}A_{i+2}$.

The first conclusion is that the number of segments of the second type is even therefore the number of segments of the first type is odd. Since the signs of the segments of the second type are paired, it follows that one extra segment of the first type appears in the sum with a positive sign, so their total contribution in S is $\frac{\sqrt{m^2+n^2}}{m}$.

Now, we evaluate the contribution in S of the segments of second type. We group them in pairs as they appear in the sequence of segments. The main point is to determine the parities of the indices i for which we have $A_i = H_l$. Since we have $H_l\left(\frac{m}{n}l, l\right)$, we consider the following $n-1$ Euclidian divisions:

$$ml = nq_l + r_l,$$

with $1 \le l \le n-1$. It follows that the points H_l have coordinates $H_l\left(q_l + \frac{r_l}{n}, l\right)$, $1 \le l \le n-1$. Their number is $n-1$, so they will contribute with $2n-2$ terms in the sum S. At any moment, a pair of such segments is determined by a point $A_s = H_l$ which is located between two points $A_{s-1} = V_k$ and $A_{s+1} = V_{k+1}$. In order to find the signs in S of the segments $A_{s-1}A_s$ and A_sA_{s+1}, we observe that $s = l + q_l + 1$. This follows from the fact that we start from $A(0,0)$ and then we count the number l of lines and the number q_l of quotients. The conclusion is that we have the signs $-A_{s-1}A_s$ when $s-1 = l+q_l$ is even and $+A_{s-1}A_s$ when $s-1 = l+q_l$ is odd.

The key point is to observe that $l+q_l$ and r_l have the same parity for all l, $1 \leq l \leq n-1$. Indeed, using congruences mod 2, we have

$$\begin{aligned} l+q_l-r_l &\equiv m\,(l+q_l-r_l) \equiv nq_l+mq_l+r_l-mr_l \\ &\equiv (m+n)\,q_l+(1-m)\,r_l = 0. \end{aligned}$$

Moreover, since $\gcd(m,n)=1$, r_l runs over all residues $1,2,\ldots,n-1$, when l takes the values $1,2,\ldots,n-1$. Therefore, we obtain that the contribution to S of two paired segments of the second type is

$$-A_{s-1}A_s + A_sA_{s+1} = \frac{\sqrt{m^2+n^2}}{m}\cdot\frac{-r_l+(n-r_l)}{n},$$

when r_l is even, and

$$A_{s-1}A_s - A_sA_{s+1} = \frac{\sqrt{m^2+n^2}}{m}\cdot\frac{r_l-(n-r_l)}{n}.$$

when r_l is odd, for all $r_l = 1,2,\ldots,n-1$.

Finally, the sum is

$$\begin{aligned} S &= \frac{\sqrt{m^2+n^2}}{m} + 2\frac{\sqrt{m^2+n^2}}{m}\left(\frac{1-2+3-\ldots+(n-2)-(n-1)}{n}\right) \\ &= \frac{\sqrt{m^2+n^2}}{m}\left(1-\frac{n-1}{n}\right) = \frac{\sqrt{m^2+n^2}}{mn}. \end{aligned}$$

The 21st BMO

The twenty-first Balkan Mathematical Olympiad for high-school students was held between May 5th and May 10th , 2004, in Pleven, Bulgaria. The number of the participating countries was 9: Albania, Bulgaria, Cyprus, Former Yugoslav Republic of Macedonia, Greece, Republic of Moldova, Romania, Turkey and Serbia-Montenegro. Teams from Kazakhstan and Yakutia (Russia) participated hors concours.

Problems

21.1. The sequence $a_0, a_1, a_2, \ldots$ of real numbers satisfies the relation

$$a_{m+n} + a_{m-n} - m + n - 1 = \frac{1}{2}\left(a_{2m} + a_{2n}\right)$$

for all non-negative integers m and n, $m \geq n$. If $a_1 = 3$ find a_{2004}.

(Cyprus)

21.2. Solve in prime numbers the equation

$$x^y - y^x = xy^2 - 19.$$

(Albania)

21.3. Let O be an interior point of an acute triangle ABC. The circles with centers the midpoints of its sides and passing through O mutually intersect the second time at the points K, L, and M different from O. Prove that O is the incenter of the triangle KLM if and only if O is the circumcenter of the triangle ABC.

(Romania)

21.4. The plane is partitioned into regions by a finite number of lines no three of which are concurrent. Two regions are called "neighbors" if the intersection of their boundaries is a segment, or halfline or a line (a point is not a segment). An integer is to be assigned to each region in such a way that:

(i) the product of the integers assigned to any two neighbors is less than their sum;

(ii) for each of the given lines, and each of the two halfplanes determined by it, the sum of the integers, assigned to all of the regions lying on this halfplane is equal to zero.

Prove that this is possible if and only if not all of the lines are parallel.

(Serbia and Montenegro)

Solutions

21.1. For $m = n$ we obtain

$$a_{2m} + a_0 - 1 = a_{2m},$$

hence $a_0 = 1$. Setting $m = n + 1$ in the given condition yields

$$a_{2n+2} - 2a_{2n+1} + a_{2n} = 2, \tag{1}$$

a condition that connects even and odd members of the sequence. For $n = 0$ we obtain $a_2 = 7$. Replacing (m, n) by $(n + 1, n - 1)$ in the given condition gives

$$a_{2n} + a_2 - 3 = \frac{1}{2}\left(a_{2n+2} + a_{2n-2}\right).$$

But $a_2 = 7$; we derive

$$a_{2n+2} - 2a_n + a_{2n-2} = 8. \tag{2}$$

This relation can be considered as a "difference equation" for the even terms. Summing up for n yields

$$a_{2n+2} - a_{2n} = 8n + 6.$$

Summing up again, we obtain

$$a_{2n+2} = 4n^2 + 10n + 7 = (2n + 2)(2n + 3) + 1.$$

Replacing in (1) leads to

$$a_{2n+1} = 4n^2 + 7n + 3 = (2n + 1)(2n + 2) + 1.$$

Therefore, the general formula is

$$a_n = n(n + 1) + 1.$$

It is easy to see that it satisfies the given condition.

Finally,

$$a_{2004} = 2004 \cdot 2005 + 1 = 4018\,021.$$

Second solution. By setting $n = 0$ in the given condition, we obtain

$$a_{2m} = 4a_m - (2m + 3)\,.$$

Taking $m = 2$ and starting with $a_0 = 1$, $a_1 = 3$, $a_2 = 7$ we get $a_4 = 21$. From (1) we have $a_3 = 13$. We can conjecture that $a_n = n^2 + n + 1$ and prove this inductively. Suppose the formula is true for all $n \leq 2m - 1$. Then

$$\begin{aligned} a_{2m} &= 4a_m - (2m + 3) \\ &= 4\left(m^2 + m + 1\right) - (2m + 3) \\ &= (2m)^2 + 2m + 1, \end{aligned}$$

and, by using (1),

$$a_{2m+1} = (2m + 1)^2 + (2m + 1) + 1,$$

as desired.

21.2. It is easy to see that $x = y$ cannot give a solution; therefore, $x \neq y$. Since y is a prime number, we have

$$x^y \equiv x \pmod y$$

Taking both sides of the equation modulo y, yields

$$x + 19 \equiv 0 \pmod y$$

which implies that $y \leq x + 19$. If $y = x + 19$, then the prime numbers x and y have different parities and it follows that $x = 2$ and $y = 21$, obviously not a solution of the equation. Therefore

$$y < x + 19. \tag{3}$$

Similarly, taking both sides of the original equation modulo x, yields

$$y - 19 \equiv 0 \pmod x\,.$$

If $y > 19$, the above congruence implies that $x \leq y - 19$, thus contradicting (3).

Consequently, $y \leq 19$, therefore we have to check a finite number of cases. It is not difficult to see that the only solutions are $(2, 3)$ and $(2, 7)$. Indeed, focusing on the two conditions $x|19 - y$ and $y|19 + x$, we have several cases:

1. $x = 2$; then $y|21$ and we have to check the pairs $(2,3)$ and $(2,7)$. Both are solutions.

2. $x = 3$; then $y|22$, so $y \in \{2,11\}$. But $3 \nmid 19 - 2 = 17$, $3 \nmid 19 - 11 = 8$.

3. $x = 5$; then $y|24$, so $y \in \{2,3\}$. But $5 \nmid 17$, $5 \nmid 16$.

4. $x = 7$; then $y|26$, so $y \in \{2,13\}$. But $7 \nmid 17$, $7 \nmid 6$.

5. $x = 11$; then $y|30$, so $y \in \{2,3,5\}$. But $11 \nmid 17, 16$ or 14.

6. $x = 13$; then $y|32$, so $y = 2$. But $13 \nmid 17$.

7. $x = 17$; then $y|36$, so $y \in \{2,3\}$. But $17 \nmid 16$. We have to check directly only the pair $(17,2)$, which is not a solution.

The conclusion is: the only solutions of the given equation are $(2,3)$ and $(2,7)$.

21.3. Let A', B', and C' be the midpoints of the sides BC, CA, and AB, respectively. It is not difficult to see that regardless of the position of the point O, the points K, L, and M are the reflections of O across the sides of triangle $A'B'C'$ (see Figure 21.1). Therefore, an homothety centered at O with ratio $\frac{1}{2}$ sends the triangle KLM to $K'L'M'$, where the points K', L', and M' are the projections of O to the sides of triangle $A'B'C'$.

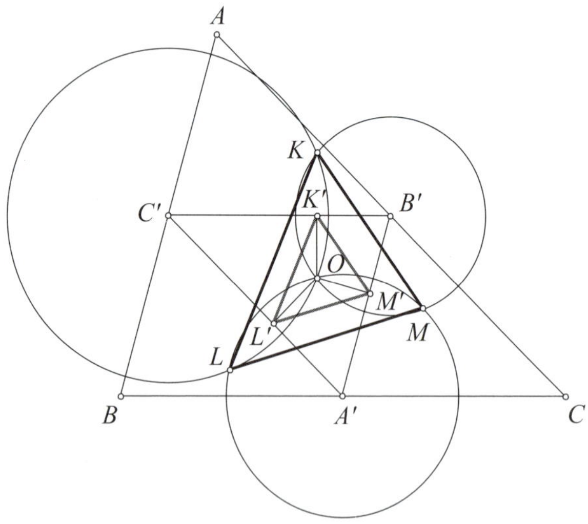

Figure 21.1

Now, suppose that O is the circumcenter of ABC. Then it is easy to see that O is also the orthocenter of $A'B'C'$. It is known that the orthocenter of a triangle is in the same time the incenter of the orthic triangle; in our case the point O is the incenter of the triangle $K'L'M'$. Indeed, we have (see Figure 21.2)

$$\angle M'K'A' = \angle L'K'A' = 90^\circ - \angle B'A'C',$$

therefore OK' is the angle bisector of $\angle M'K'L'$. Similarly, OM' and OL' are also angle bisectors, hence O is the incenter of triangle $K'L'M'$.

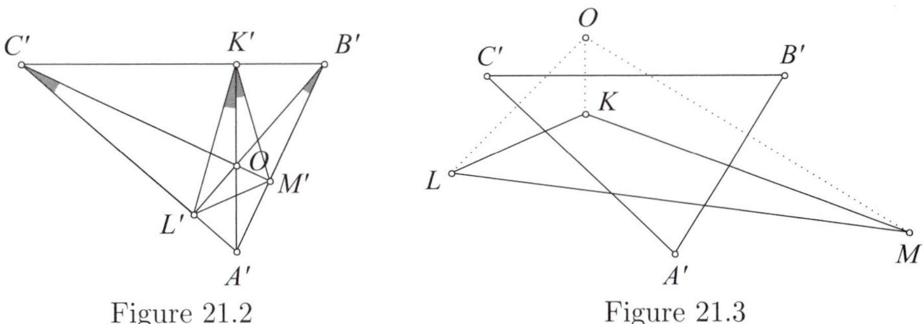

Figure 21.2 Figure 21.3

Conversely, suppose that O is the incenter of KLM. Then is the incenter of $K'L'M'$ as well (recall that the points K', L', and M' are the projections of O to the sides of triangle $A'B'C'$). We observe that O lies in the interior of the triangle $A'B'C'$. Indeed, if not, then O is an exterior point of the triangle $K'L'M'$ (see Figure 21.3), contradicting thus our assumption. It is not difficult to see that O is the orthocenter of $A'B'C'$ and hence the circumcenter of ABC.

Second solution. If O is the circumcenter of triangle ABC, then $\angle A'C'O = \angle A'B'O = 90^\circ - \angle A$. But

$$\angle A'C'O = \frac{\angle LC'O}{2} = \angle LKO \text{ and } \angle A'B'O = \frac{\angle MB'O}{2} = \angle MKO.$$

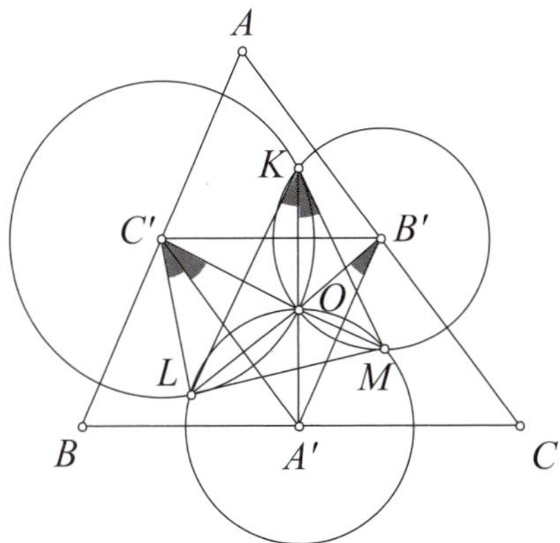

Figure 21.4

hence OK is the angle bisector of $\angle LKM$ (see Figure 21.4). Similarly, OM and OL are angle bisectors, therefore O is the incenter of triangle KLM.

Conversely, if O is the incenter of triangle KLM, then $\angle LKO = \angle MKO$. But $\angle LKO = \angle A'C'O$ and $\angle MKO = \angle A'B'O$, therefore $\angle A'C'O = \angle A'B'O$. Similarly, $\angle B'A'O = \angle B'C'O$ and $\angle C'A'O = \angle C'B'O$. Adding up yields

$$\angle C'A'O + \angle A'C'O + \angle B'C'O = 90^\circ.$$

It follows that $A'O$ is perpendicular to $B'C'$ and hence, to BC. Then $A'O$ is the perpendicular bisector of the line segment BC. In the same way we prove that $B'O$ and $C'O$ are the perpendicular bisectors of AC and AB, therefore O is the circumcenter of triangle ABC.

Observation. Let D, E, and F be the feet of the altitudes of the triangle ABC, and let ω be the center of the nine point circle. It is known (see Appendix) that ω is the midpoint of the line segment OH, where O and H are the circumcenter and the orthocenter of ABC. We can see that a symmetry across ω sends H to O and the orthic triangle DEF to the triangle KLM (see Figure 21.5).

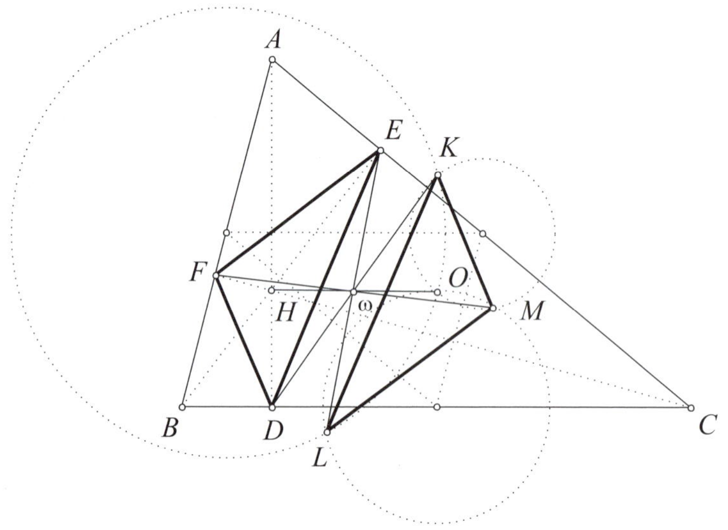

Figure 21.5

21.4. If all the lines are parallel, it is easy to check that all numbers must be equal to zero and then the condition (i) is not satisfied.

So, we assume that there are at least two nonparallel lines. For each line, we assign to the halfplanes which it defines either $+$ or $-$. We color each point of the plane with the product of the signs assigned to the halfplanes containing that point (see Figure 21.6).

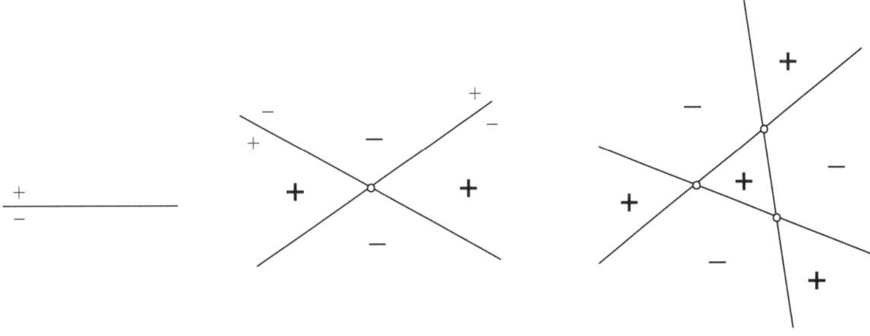

Figure 21.6

In this way, all the points of a region have the same color, which is the color of the region.

Since no three of the given lines are concurrent, any point of intersection between two lines is adjacent to four regions. In each of these regions we write around the point $+1$ or -1, corresponding to the color of the region (see Figure 21.7). Finally, we assign to each region the sum of the numbers written in its corners, in the interior of that region (see Figure 21.8).

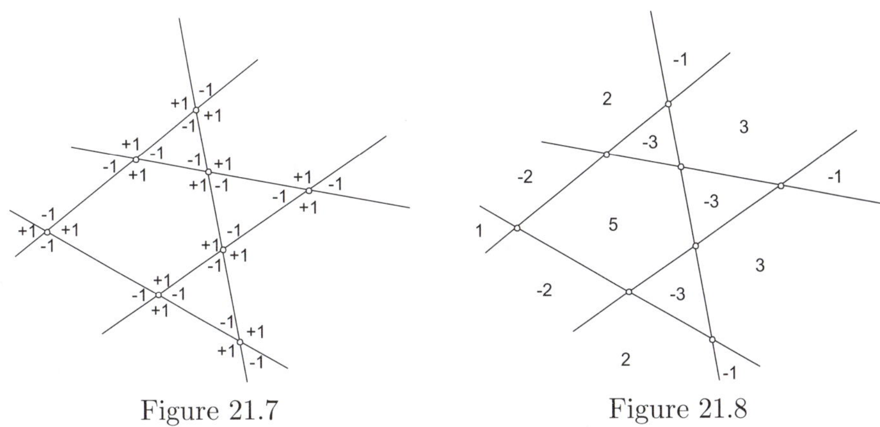

Figure 21.7 Figure 21.8

It is easy to check that the condition are satisfied. For (i) we have $ab \leq a < a + b$, since either $a < 0 < b$ or $b < 0 < a$ and a, b are integers (here a and b are numbers assigned to two neighboring regions). For (ii), observe that the sum of the numbers assigned to the regions contained in a halfplane H is equal to the sum of ± 1 written around all intersection points belonging to H. Since the sum of ± 1 written around each intersection point is obviously zero, we are done.

The 22nd BMO

The twentysecond Balkan Mathematical Olympiad for high-school students was held between May 4th and May 10th , 2005, in Iaşi, Romania. The number of the participating countries was 9: Albania, Bulgaria, Cyprus, Former Yugoslav Republic of Macedonia, Greece, Republic of Moldova, Romania, Turkey and Serbia-Montenegro. Teams from Hungary, United Kingdom, Kazakhstan and Yakutia (Russia) participated hors concours.

Problems

22.1. Let ABC be an acute-angled triangle whose inscribed circle touches AB and AC at D and E, respectively. Let X and Y be the points of intersection of the bisectors of the angles $\angle ACB$ and $\angle ABC$ with the line DE and let Z be the midpoint of the side BC. Prove that the triangle XYZ is equilateral if and only if $\angle A = 60°$.

(Bulgaria)

22.2. Find all primes p such that $p^2 - p + 1$ is a perfect cube.

(Albania)

22.3. Let a, b, c be positive real numbers. Prove the inequality

$$\frac{a^2}{b} + \frac{b^2}{c} + \frac{c^2}{a} \geq a + b + c + \frac{4(a-b)^2}{a+b+c}.$$

When does the equality hold?

(Serbia and Montenegro)

22.4. Let $n \geq 2$ be an integer. Let S be a subset of $\{1, 2, \ldots, n\}$ such that S neither contains two elements one of which divides the other, nor contains two elements which are coprime. What is the maximum possible number of elements of such a set S?

(Romania)

Solutions

22.1. Let I denote the incenter of triangle ABC. We claim that the quadrilateral $BIXD$ is cyclic. Indeed, observe that the triangle ADE is isosceles, hence $\angle AED = \angle ADE = 90^\circ - \frac{1}{2}A$. It follows that $\angle DEC = 90^\circ + \frac{1}{2}A$ and then

$$\angle DXI = \angle DEC + \angle ECX = 90^\circ + \frac{A}{2} + \frac{C}{2}.$$

Since $\angle DBI = \frac{1}{2}B$, the angles $\angle DXI$ and $\angle DBI$ add up to 180°, thus proving our claim.

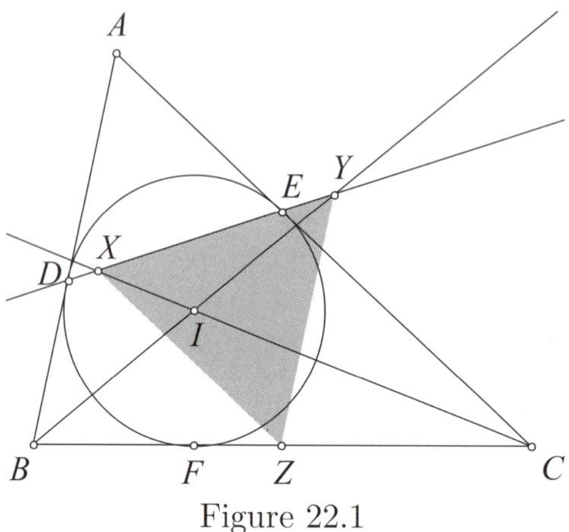

Figure 22.1

Now, since ID is perpendicular to AB it follows that $\angle BXC = 90^\circ$ and hence $XZ = BZ = CZ$. In a similar way we deduce that $YZ = BZ = CZ$, so $XZ = YZ$, that is, the triangle XYZ is always isosceles.

Finally, since $XZ = CZ$, we have $\angle ZXC = \angle XCZ = \angle ACX$, hence XZ is parallel to AC. Similarly, YZ is parallel to AB. This implies that the angles $\angle XZY$ and $\angle BAC$ are equal, and the conclusion is straightforward.

Second solution. We use complex numbers, considering the incircle of triangle ABC to be the unit circle.

Lemma. Let A and B be two points on the unit circle, and let C be the intersection of the tangents to the circle passing through A and B. If a, b and c are the complex numbers corresponding to the points A, B and C, then

$$c = \frac{2ab}{a+b}.$$

Proof. Since $\angle OAC = 90^\circ$, the number $\frac{c-a}{a}$ is purely imaginary, therefore

$$\frac{c-a}{a} = -\frac{\overline{c} - \overline{a}}{\overline{a}}.$$

Using that $a\overline{a} = 1$, the equality above simplifies to

$$c + a^2\overline{c} = 2a.$$

Similarly, we have

$$c + b^2\overline{c} = 2b.$$

Solving for c we obtain the desired equality.

Returning to the problem, let $a, b, c, ...$ denote the complex numbers that correspond to the points A, B, C, etc. Furthermore, suppose that the incircle touches the side BC at F. We can assume with no loss of generality that $f = 1$. Using the lemma, we obtain

$$b = \frac{2d}{1+d}, \qquad c = \frac{2e}{1+e},$$

and

$$a = \frac{2de}{d+e}.$$

Since X lies on the line DE, we have

$$\frac{x-d}{x-e} = \frac{\overline{x}-\overline{d}}{\overline{x}-\overline{e}},$$

and using $d\overline{d} = e\overline{e} = 1$, the latter simplifies to

$$x + \overline{x}de = d + e.$$

Since X lies on the line CI, we also have

$$\frac{x}{c} = \frac{\overline{x}}{\overline{c}},$$

and this yields

$$x = \overline{x}e.$$

We deduce

$$x = \frac{d+e}{1+d},$$

and, in a similar way,

$$y = \frac{d+e}{1+e}.$$

Now, Z is the midpoint of the line segment BC, therefore

$$z = \frac{b+c}{2} = \frac{d}{1+d} + \frac{e}{1+e},$$

and then

$$\frac{x-z}{y-z}=\frac{\frac{d+e}{1+d}-\frac{d}{1+d}-\frac{e}{1+e}}{\frac{d+e}{1+e}-\frac{d}{1+d}-\frac{e}{1+e}}=-\frac{e}{d}.$$

On the other hand,

$$\frac{e-a}{d-a}=\frac{e-\frac{2de}{d+e}}{d-\frac{2de}{d+e}}=-\frac{e}{d}.$$

Because

$$\frac{x-z}{y-z}=\frac{e-a}{d-a},$$

we deduce that the triangles XZY and EAD are similar. Obviously ABC is an isosceles triangle, therefore it is equilateral if and only if $A=60°$.

22.2. Let

$$p^2-p+1=q^3,$$

where q is a positive integer. Rewrite the latter as

$$p(p-1)=(q-1)\left(q^2+q+1\right).$$

It follows that p divides $q-1$ or q^2+q+1.
Suppose that p divides $q-1$.Then we have $p<q<q^2+q+1$ and $p-1<q-1$, which leads to a contradiction.

Consequently, it follows that p divides q^2+q+1, say

$$q^2+q+1=np, \tag{1}$$

for some positive integer n. We obtain

$$p(p-1)=(q-1)\,np,$$

hence

$$p=1+n(q-1).$$

Replacing in (1) yields

$$q^2+q\left(1-n^2\right)+n^2-n+1=0.$$

Consider this as a quadratic in q; then its discriminant Δ must be a square. But

$$\Delta=n^4-6n^2+4n-3,$$

and it is not difficult to see that

$$\left(n^2-3\right)^2\le\Delta<\left(n^2-2\right)^2.$$

It follows that $\Delta = \left(n^2 - 3\right)^2$ and we deduce that $n = 3$, $q = 7$ and $p = 19$.

22.3. Observe that

$$\frac{a^2}{b} - 2a + b = \frac{(a-b)^2}{b},$$

hence the given inequality can be rewritten as

$$\frac{(a-b)^2}{b} + \frac{(b-c)^2}{c} + \frac{(c-a)^2}{a} \geq \frac{4(a-b)^2}{a+b+c}.$$

The Cauchy-Schwartz inequality yields

$$(b+c+a)\left(\frac{(a-b)^2}{b} + \frac{(b-c)^2}{c} + \frac{(c-a)^2}{a}\right) \geq (|a-b| + |b-c| + |c-a|)^2. \tag{2}$$

On the other hand, we have

$$|b-c| + |c-a| \geq |(b-c) + (c-a)| = |b-a| = |a-b|, \tag{3}$$

hence

$$(|a-b| + |b-c| + |c-a|)^2 \geq (2|a-b|)^2 = 4(a-b)^2.$$

Finally, let us analyze the cases when equality holds. We have equality in (2) if and only if

$$\frac{|a-b|}{b} = \frac{|b-c|}{c} = \frac{|c-a|}{a},$$

and (3) becomes an equality if and only if $b-c$ and $c-a$ have the same sign. Therefore we have either

$$\frac{a-b}{b} = \frac{b-c}{c} = \frac{c-a}{a} = \frac{a-b+b-c+c-a}{b+c+a} = 0,$$

leading to $a = b = c$, or

$$\frac{b-a}{b} = \frac{b-c}{c} = \frac{c-a}{a}.$$

The latter can be written as

$$1 - \frac{a}{b} = \frac{b}{c} - 1 = \frac{c}{a} - 1,$$

hence

$$\frac{b}{c} = \frac{c}{a}.$$

Denote by x the common value of the two fractions. Then $x^2 = \frac{b}{a}$ and we obtain

$$1 - \frac{1}{x^2} = x - 1,$$

or

$$x^3 - 2x^2 + 1 = 0.$$

The latter factors as

$$(x-1)\left(x^2 - x - 1\right) = 0,$$

hence $x = 1$, leading again to $a = b = c$, or $x = \frac{1+\sqrt{5}}{2}$ (since a, b, c are positive numbers, we can discard the negative root of the equation).

We conclude that the equality holds when either $a = b = c$, or $c = a\varphi$, $b = a\varphi^2$, where $\varphi = \frac{1+\sqrt{5}}{2}$.

22.4. The natural approach is to try to build such a set. Chosing all numbers greater than $\frac{n}{2}$ solves the first condition, because if a and b are distinct integers greater than $\frac{n}{2}$ and less than or equal to n, then $\frac{a}{b}$ and $\frac{b}{a}$ cannot be integers, hence none of the numbers divides the other. To fulfill the second condition it suffices to consider all the numbers even. Thus, an example of a set with the desired properties is

$$S = \left\{ k \mid \frac{n}{2} < k \leq n,\ k \text{ even} \right\},$$

and it is not difficult to see that the number of elements of this set equals $\left\lfloor \frac{n+2}{4} \right\rfloor$.

We claim that this is the maximum number. Indeed, let S be a set with the required properties and let a be its least element. If $a \leq \frac{n}{2}$, replace a by $2a$. Clearly, the new set still has the required properties. Repeat this transformation until all the elements are greater than $\frac{n}{2}$. Now, since there are no coprime numbers in the set, there are no consecutive integers either. It is not difficult to see that such a set cannot have more than $\left\lfloor \frac{n+2}{4} \right\rfloor$ elements.

The 23rd BMO

The twentythird Balkan Mathematical Olympiad for high-school students was held between April 27th and May 3rd, 2006, in Agros, Cyprus. The number of the participating countries was 9: Albania, Bulgaria, Cyprus, Former Yugoslav Republic of Macedonia, Greece, Republic of Moldova, Romania, Turkey and Serbia-Montenegro. Teams from Hungary, United Kingdom, Kazakhstan and Yakutia (Russia) participated hors concours.

Problems

23.1. Let a, b, c be positive numbers. Prove the inequality

$$\frac{1}{a(1+b)}+\frac{1}{b(1+c)}+\frac{1}{c(1+a)} \geq \frac{3}{1+abc}.$$

(Greece)

23.2. Let ABC be a triangle. A line l intersects the sides AB, AC and the extension of BC beyond C at the points D, F, and E, respectively. The lines through A, B, C which are parallel to l meet again the circumcircle of the triangle at points A_1, B_1, and C_1, respectively. Prove that the lines A_1E, B_1F, and C_1D are concurrent.

(Greece)

23.3. Determine all triples (m, n, p) of positive rational numbers such that the numbers$m+\dfrac{1}{np}$, $n+\dfrac{1}{pm}$, $p+\dfrac{1}{mn}$ are integers.

(Romania)

23.4. Given a positive integer m, consider the sequence (a_n) of positive integers defined by the initial term $a_0 = a$ and the recurrent relation

$$a_{n+1} = \begin{cases} \frac{a_n}{2}, & \text{if } a_n \text{ is even;} \\ a_n + m, & \text{if } a_n \text{ is odd.} \end{cases}$$

Find all values of a for which the sequence is periodic.

(Bulgaria)

Solutions

23.1. Rewrite the inequality as

$$\frac{1+abc}{a\,(1+b)}+\frac{1+abc}{b\,(1+c)}+\frac{1+abc}{c\,(1+a)}\geq 3,$$

and add 1 to each of the terms of the left hand side. This yields

$$\frac{1+abc+a+ab}{a\,(1+b)}+\frac{1+abc+b+bc}{b\,(1+c)}+\frac{1+abc+c+ca}{c\,(1+a)}\geq 6,$$

or, equivalently,

$$\frac{1+a}{a\,(1+b)}+\frac{b(1+c)}{1+b}+\frac{1+b}{b\,(1+c)}+\frac{c\,(a+1)}{1+c}+\frac{1+c}{c\,(1+a)}+\frac{a\,(b+1)}{1+a}\geq 6.$$

Since the product of the six terms of the left hand side equals 1, the inequality follows from the $AM-GM$ inequality.

Second solution. Let $abc=t^3$ and let x,y,z be positive numbers such that

$$a=t\frac{x}{y},b=t\frac{y}{z},c=t\frac{z}{x}.$$

Replacing in the original inequality yields

$$\frac{yz}{t^2xy+tzx}+\frac{zx}{t^2yz+txy}+\frac{xy}{t^2zx+tyz}\geq\frac{3}{1+t^3},$$

and denoting $yz=u,zx=v,xy=w$ we obtain the equivalent inequality

$$\frac{u}{t^2v+tw}+\frac{v}{t^2w+tu}+\frac{w}{t^2u+tv}\geq\frac{3}{1+t^3}.$$

The latter can be proved using the Cauchy-Schwartz inequality. We have

$$\left(\sum\left(t^2uv+twu\right)\right)\left(\sum\frac{u^2}{t^2uv+twu}\right)\geq(u+v+w)^2,$$

therefore

$$\frac{u}{t^2v+tw}+\frac{v}{t^2w+tu}+\frac{w}{t^2u+tv}\geq\frac{(u+v+w)^2}{(t^2+t)\,(uv+vw+wu)}.$$

Since

$$(u+v+w)^2\geq 3\,(uv+vw+wu)\,,$$

all we have to prove is

$$\frac{1}{t^2+t} \geq \frac{1}{1+t^3}.$$

A short computation shows that this is equivalent to the obvious

$$(t-1)^2(t+1) \geq 0.$$

Observations. 1. A proof by "brute force" is also possible. Clearing denominators and rearranging terms lead to the equivalent inequality

$$ab(b+1)(ac-1)^2 + bc(c+1)(ab-1)^2 + ca(a+1)(bc-1)^2 \geq 0,$$

obviously true.

2. Actually, the inequality is a disguise of an old an famous inequality, proposed in 1988 in the Russian magazine "Kvant", in honour of the centennial of the American Mathematical Society:

$$3+A+M+S+\frac{1}{A}+\frac{1}{M}+\frac{1}{S}+\frac{A}{M}+\frac{M}{S}+\frac{S}{A} \geq \frac{3(A+1)(M+1)(S+1)}{1+AMS}.$$

23.2. Suppose that the line B_1F intersects the circumcircle at P (see Figure 23.1). We claim that the points C_1, D and P are collinear, hence C_1D passes through P. Similarly, we obtain that A_1E passes through P as well, thus proving that the three lines are concurrent.

To prove the claim, observe that the quadrilateral $ADFP$ is cyclic. This is because ABB_1P is cyclic and DF is parallel to BB_1, hence the angles of $ADFP$ are equal to the angles of ABB_1P.

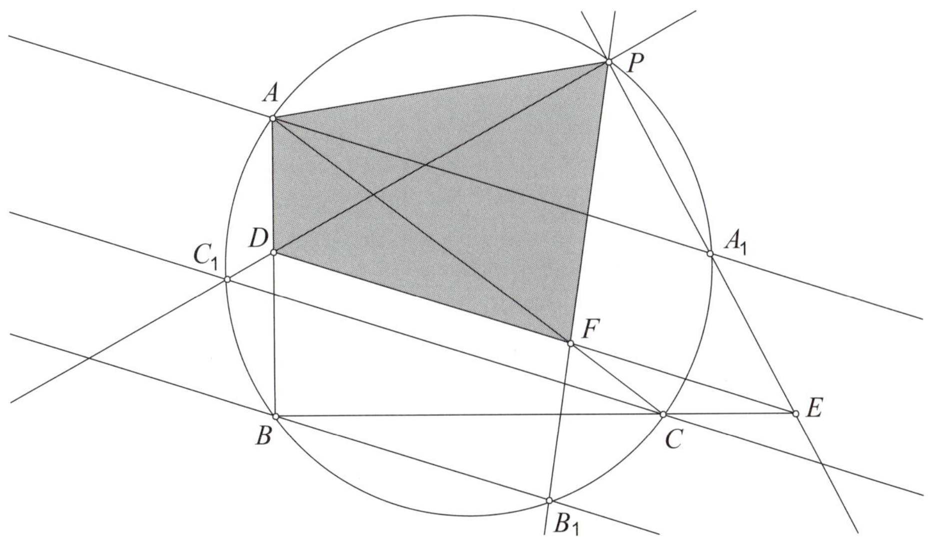

Figure 23.1

Therefore, $\angle DPF = \angle BAC$. But $\angle BAC = \angle C_1PB_1$, because the arcs B_1C and BC_1 are equal. It follows that $\angle DPF = \angle C_1PF$, that is, the points C_1, D, and P are collinear.

23.3. Multiplying the three numbers, we obtain that

$$\frac{(mnp+1)^3}{m^2n^2p^2}$$

is an integer as well. Let $x = mnp$ and suppose that

$$\frac{(x+1)^3}{x^2} = k,$$

for some integer k. Rewrite the latter as

$$x^3 + (3-k)\,x^2 + 3x + 1 = 0.$$

Now, x is a rational number, say $x = \frac{a}{b}$, with $\gcd(a,b) = 1$. We obtain

$$a^3 + (3-k)\,a^2b + 3ab^2 + b^3 = 0.$$

It follows that b divides a, and a divides b, hence $x = 1$.

Finally, we have to find the positive rational numbers m, n, p such that $mnp = 1$ and $2m$, $2n$, and $2p$ are integers. It is not difficult to see that the triples satisfying these conditions are $(1,1,1)$, $\left(2,1,\frac{1}{2}\right)$, $\left(4,\frac{1}{2},\frac{1}{2}\right)$ and all their permutations.

23.4. Note that if m is even, the sequence cannot be periodic. Indeed, some term has to be odd and, if m were even, from that term on the sequence is increasing. Suppose that m is odd and that the period of the sequence is $a_0, a_1, \ldots, a_k$ (that is, $a_{k+1} = a_0$, etc.) Denote by A the set of the possible values of the initial term $a_0 = a$.

We can assume with no loss of generality that a_0 is the smallest element from the set $\{a_0, a_1, \ldots, a_k\}$. Then a_0 must be odd (otherwise $a_1 = \frac{a_0}{2} < a_0$) and $a_0 \le a_2 = \frac{a_0+m}{2}$, hence $a_0 \le m$. If for some i we have $a_i \le m$, then either $a_{i+1} = \frac{a_i}{2} \le m$, or $a_{i+1} = a_i + m \le 2m$ and a_{i+1} is even, but in the second case $a_{i+2} = \frac{a_{i+1}}{2} \le m$. We deduce that the elements of A which are greater than m must be even and less than or equal to $2m$.

We claim that $A = \{1, 2, \ldots, m-1, m, m+1, m+3, \ldots, 2m-2, 2m\}$. To see this, consider an oriented graph whose vertices are the elements of A and whose edges are as follows: from x to $\frac{x}{2}$ if x is even and from x to $x+m$ if x is odd. Obviously, the out-degree of each vertex equals 1. On the other hand, if $x \le m$, the only in-edge to x comes from $2x$, while if $x > m$, the only in-edge comes from $x - m$. Thus, the in-degree of each vertex equals 1 as well. This means that the graph can be partitioned into cycles, thus proving the claim.

The 24th BMO

The twenty-fourth Balkan Mathematical Olympiad for high-school students was held between April 26th and May 2nd, 2007, in Rhodes, Greece. The number of the participating countries was 14: 10 member countries (Albania, Bulgaria, Cyprus, Former Yugoslav Republic of Macedonia, Greece, Republic of Moldova, Montenegro, Romania, Serbia and Turkey) and 4 invited countries (Azerbaijan, Italy, Kazakhstan and United Kingdom).

Problems

24.1. Let $ABCD$ a convex quadrilateral with $AB = BC = CD$, with AC not equal to BD and let E be the intersection point of its diagonals. Prove that $AE = DE$ if and only if $\angle BAD + \angle ADC = 120°$.

(Albania)

24.2. Find all functions $f : \mathbb{R} \to \mathbb{R}$ such that

$$f(f(x)+y) = f(f(x)-y) + 4f(x)y,$$

for any real numbers x, y.

(Bulgaria)

24.3. Find all positive integers n such that there exists a permutation σ of the set $\{1, 2, \ldots, n\}$ for which

$$\sqrt{\sigma(1) + \sqrt{\sigma(2) + \sqrt{\ldots + \sqrt{\sigma(n)}}}}$$

is a rational number.

(Serbia)

24.4. For a given positive integer $n > 2$, let C_1, C_2, C_3 be the boundaries of three convex $n-$gons in the plane, such that $C_1 \cap C_2$, $C_2 \cap C_3$, and $C_3 \cap C_1$ are finite. Find the maximum number of points of the set $C_1 \cap C_2 \cap C_3$.

(Turkey)

Solutions

24.1. We will use the following

Lemma. Let $MNPQ$ be a convex quadrilateral such that $MN = NP$ and $\angle MQN = \angle NQP$. Prove that if Q does not lie on the perpendicular bisector of the line segment MP, then $MNPQ$ is cyclic.

Proof. Assume the contrary and consider the case when Q lies in the interior of the circumcircle of the triangle MNP (see Figure 24.1). Extend NQ until it meets the circumcircle at R. Since $MN = NP$, the angles MRN and NRP are congruent, and we deduce that the triangles MRQ and PRQ are congruent, as well. It follows that $QP = QM$, hence Q lies on the perpendicular bisector of MP, a contradiction.

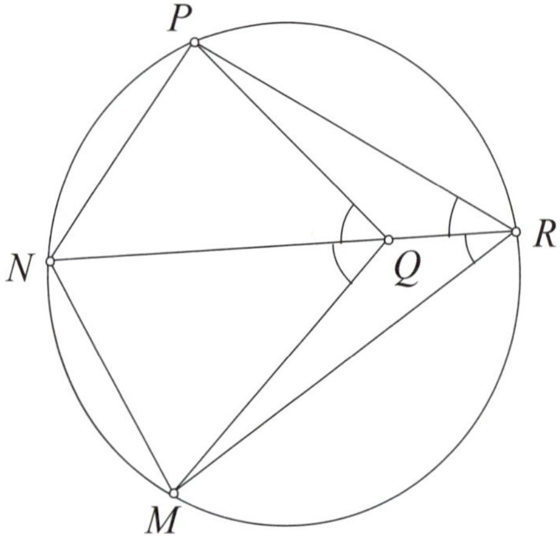

Figure 24.1

The case when Q lies in the exterior of the circumcircle can be treated in a similar manner.

Returning to our problem, let us consider the point F such that $ACDF$ is a parallelogram (see Figure 24.2).

Observe that $AE = ED$ is equivalent to $\angle CAD = \angle ADB$ and, furthermore, to $\angle ADB = \angle ADF$, since AC is parallel to FD.

It is not difficult to see that D cannot lie on the perpendicular bisector of FB, because if it were, we would have $DB = DF$, and since $AC = DF$, that would imply $BD = AC$, contradicting the hypothesis.

Applying the above lemma, we deduce that $AE = ED$ if and only if $ABDF$ is a cyclic quadrilateral.

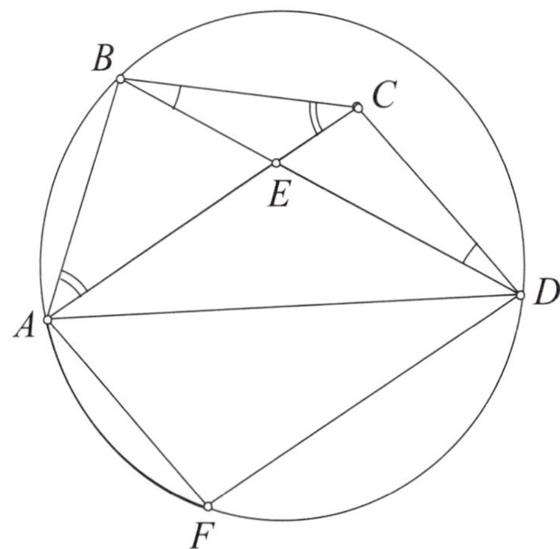

Figure 24.2

Let $s = \angle BAD + \angle ADC$, and observe that

$$\angle FAB = \angle BAD + \angle FAD = \angle BAD + \angle ADC = s.$$

Also,

$$\angle FDB = \angle FDA + \angle ADE = \angle DAE + \angle ADE.$$

But

$$\begin{aligned}\angle DAE + \angle ADE &= 180^\circ - \angle AED = 180^\circ - \angle BEC \\ &= \angle EBC + \angle ECB = \angle BDC + \angle CAB \\ &= s - (\angle DAE + \angle ADE),\end{aligned}$$

hence

$$\angle DAE + \angle ADE = \frac{s}{2}.$$

Now, $ACDF$ is a cyclic quad if and only if $\angle FAB + \angle FDB = 180^\circ$. Since

$$\angle FAB + \angle FDB = s + \frac{s}{2} = \frac{3s}{2},$$

this is equivalent to $s = 120^\circ$, as desired.

Second solution. Suppose that $AE = DE$ and examine the triangles ABE and DCE. They have two pairs of congruent sides and $\angle AEB = \angle CED$. Applying the sine law in these triangles yields

$$\frac{AE}{\sin \angle ABE} = \frac{AB}{\sin \angle AEB} = \frac{CD}{\sin \angle CED} = \frac{ED}{\sin \angle ACD}.$$

It follows that $\sin \angle ABE = \sin \angle ACD$, hence these two angles are either equal or supplementary.

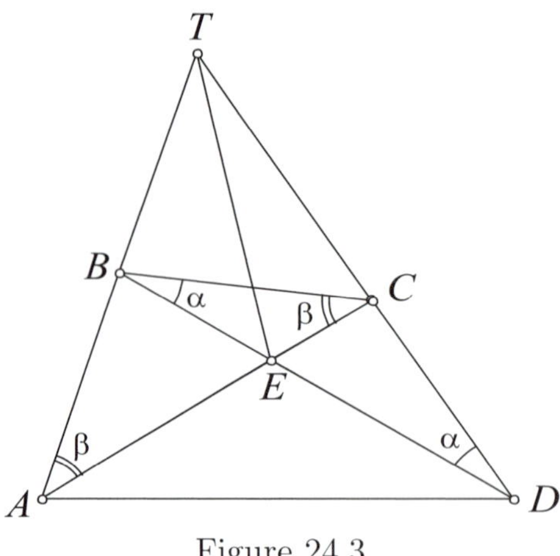

Figure 24.3

If $\angle ABE = \angle ACD$, then

$$180^\circ - (\alpha + 2\beta) = 180^\circ - (2\alpha + \beta),$$

hence $\alpha = \beta$. This obviously implies $AC = BD$, which is not true.

It remains that $\angle ABE + \angle ACD = 180^\circ$, and we easily derive that $\alpha + \beta = 60^\circ$.

Since

$$\angle EAD + \angle EDA = 180^\circ - \angle AED = 180^\circ - \angle BEC = \alpha + \beta,$$

it follows that

$$\angle BAD + ADC = 2(\alpha + \beta) = 120^\circ.$$

Conversely, suppose that $\angle BAD + ADC = 120^\circ$, and let $T = AB \cap CD$. It follows that $\angle ATD = 60^\circ$, and $\alpha + \beta = 60^\circ$, as well. Hence $\angle ATD + \angle BEC = 180^\circ$, that is, the quadrilateral $TBEC$ is cyclic.

Then $\angle BTE = \angle BCE = \angle BAE$, so the triangle TAE is isosceles. It follows that $TE = AE$. Similarly we obtain $TE = DE$, hence $AE = DE$.

24.2. Obviously, $f = 0$ is a solution of the functional equation. Let x_0 be a real number such that $f(x_0) \neq 0$. Then

$$f(f(x_0) + y) = f(f(x_0) - y) + 4f(x_0)y,$$

or, equivalently,

$$8f(x_0)y = 2(f(f(x_0) + y) - f(f(x_0) - y)).$$

Thus, any real number can be written as $2(f(a) - f(b))$, for some real numbers a, b.

Now, let $y = f(x)$ in the original equation. This gives

$$f(2f(x)) = f(0) + 4f^2(x),$$

for all real x. Finally, replace y by $f(x) - 2f(y)$. We obtain

$$\begin{aligned} f(2(f(x) - f(y))) &= f(2f(y)) + 4f(x)(f(x) - 2f(y)) \\ &= f(0) + 4f^2(y) + 4f^2(x) - 8f(x)f(y) \\ &= f(0) + 4(f(x) - f(y))^2. \end{aligned}$$

Using the previous observation, we deduce that $f(x) = f(0) + x^2$, for all real x, and it is easy to check that this is indeed a solution of the functional equation.

Second solution. It is not difficult to guess that one solution of the functional equation (apart from the obvious $f = 0$) is $f(x) = x^2$. Therefore, considering the function $g(x) = f(x) - x^2$ might be a good idea.

Indeed, replacing in the given equation gives

$$g(f(x) + y) = g(f(x) - y),$$

and then, replacing y by $y - f(x)$,

$$g(y) = g(2f(x) - y),$$

for all real x, y. This implies

$$g(0) = g(2f(x)),$$

for all real x, then

$$g(0) = g(2f(y)) = g(2f(x) - 2f(y)).$$

As we saw in the previous solution, if $f \neq 0$, any real number t can be written as $t = 2f(x) - 2f(y)$, for some x, y, hence

$$g(t) = g(0),$$

for all t, leading to $f(x) = f(0) + x^2$, for all real x.

24.3. Let

$$E_k = \sqrt{\sigma(k) + \sqrt{\sigma(k+1) + \sqrt{\ldots + \sqrt{\sigma(n)}}}},$$

where $k \in \{1, 2, \ldots, n\}$. We have

$$E_{k-1} = \sqrt{\sigma(k-1) + E_k},$$

for $k \geq 2$. Since E_1 is a rational number, the above recursive equation shows that all E_k's are rational. Moreover, since the square root of a positive integer is either an integer or an irrational number, it follows that E_n is an integer, and again, using the recursive relation, that all E_k's are positive integers.

Now, we obviously have

$$E_k \leq \sqrt{n + \sqrt{n + \sqrt{\ldots + \sqrt{n}}}},$$

where in right hand side (like in E_k) there are $n - k + 1$ radicals, and a simple inductive argument shows that

$$\sqrt{n + \sqrt{n + \sqrt{\ldots + \sqrt{n}}}} < \sqrt{n} + 1.$$

Let $\lfloor \sqrt{n} \rfloor = p$, hence $\sqrt{n} + 1 < p + 2$. If $p > 1$, then $p^2 - 1$ is not a square, hence $p^2 - 1 = \sigma(m)$, for some $m < n$ (recall that $\sqrt{\sigma(n)} = E_n$ is an integer). We have $p \leq E_m < p + 2$, so, if $E_m > p$, then $E_m = p + 1$, and

$$E_{m+1} = E_m^2 - \sigma(m) = (p+1)^2 - (p^2 - 1) = 2p + 2 > p + 2,$$

a contradiction.

The only other case is $E_m = p$, but this implies $m = n - 1$ and $\sigma(n) = 1$. Considering $s < n - 1$ such that $\sigma(s) = p^2$, we again reach a contradiction, since E_s cannot be an integer.

We are left with the case $p = 1$, and it is not difficult to see that we only have two solutions: the trivial $n = 1$ and $n = 3$, since $\sqrt{2 + \sqrt{3 + \sqrt{1}}}$ is a rational number.

24.4. We claim that the answer is $\left\lfloor \frac{3n}{2} \right\rfloor$. Let P_k be the polygon with the boundary C_k, and let $P = P_1 \cap P_2 \cap P_3$.

Obviously, P is a convex polygon and each of its sides is included in a side of exactly one of the P_k's.

For $k = 1, 2, 3$, let a_k be the number of sides of P included in sides of P_k. If P has m vertices, then we obviously have

$$a_1 + a_2 + a_3 = m.$$

We will count twice the sides of P_k. On one hand, P_k has n sides. On the other hand, there are a_k sides of P_k which include sides of P (and along with these, $2a_k$ vertices of P) and for each of the $m-2a_k$ remaining vertices of P there exists a side of P_k containing the respective vertex.

It follows that

$$a_k + (m - 2a_k) \leq n,$$

hence

$$m - a_k \leq n.$$

Adding up for $k = 1, 2, 3$ yields

$$3m - (a_1 + a_2 + a_3) \leq 3n \Longleftrightarrow 2m \leq 3n \Longleftrightarrow m \leq \left\lfloor \frac{3n}{2} \right\rfloor .$$

We have to show that this bound can be reached. If n is even, say $n = 2t$, then $m = 3t$; label clockwise the sides of P with $1, 2, \ldots, 3t$. Then P_k will be the polygon determined by the t sides of P whose labels are congruent to k modulo 3, and another t sides, obtained by drawing appropriate lines through the t remained vertices of P.

A similar construction can be performed when n is odd, only that one of the P_k's will contain two consecutive sides of P (see Figure 24.4).

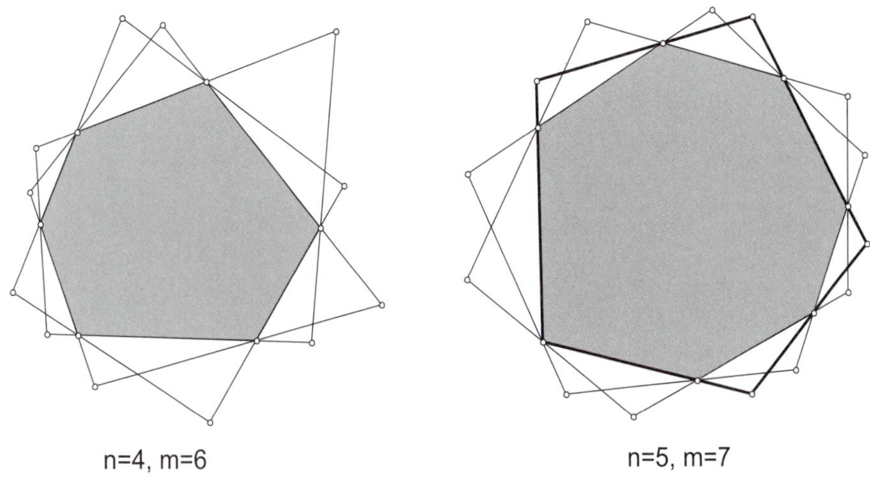

Figure 24.4

The 25th BMO

The twenty-fifth Balkan Mathematical Olympiad for high-school students was held between May 4th and May 10th, 2008, in the historical city of Ohrid, F.Y.R of Macedonia. The number of participating countries was 18: 11 member countries (Albania, Bosnia and Herzegovina, Bulgaria, Cyprus, Former Yugoslav Republic of Macedonia, Greece, Republic of Moldova, Republic of Montenegro, Romania, Serbia and Turkey) and 7 invited countries (Azerbaijan,France, Italy, Kazakhstan, Tajikistan, Turkmenistan and United Kingdom.) Additionally, a second team from Macedonia took part as Macedonia B.

Problems

25.1. Let ABC be a scalene acute angled triangle such that $AC > BC$. Let O be its circumcenter, H its orthocenter and F the foot of the altitude from C. Let P be a point on AB, other than A, so that $AF = PF$ and M the midpoint of the side AC. The line segments BC and HP meet at X, the lines OM and FX meet at Y, and OF and AC meet at Z. Prove that the points F, M, Y, Z lie on a circle.

(Cyprus)

25.2. Does there exist a sequence $a_1, a_2, \ldots, a_n, \ldots$ of positive real numbers which satisfies both of the following conditions:

(i) $a_1 + a_2 + \cdots + a_n \leq n^2$, for every positive integer n,

(ii) $\frac{1}{a_1} + \frac{1}{a_2} + \cdots + \frac{1}{a_n} \leq 2008$, for every positive integer n?

(Bulgaria)

25.3. Let n be a positive integer. The rectangle $ABCD$ with side lengths $AB = 90n + 1$ and $BC = 90n + 5$ is partitioned into unit squares with sides parallel to the sides of $ABCD$. Let S be the set of all points which are vertices of these unit squares. Prove that the number of distinct lines passing through at least two points of S is divisible by 4.

(Bulgaria)

25.4. Let c be a positive integer. The sequence $a_1, a_2, \ldots, a_n, \ldots$ is defined by $a_1 = c$ and $a_{n+1} = a_n^2 + a_n + c^3$, for every positive integer n. Find all values of c for which there exist some integers $k \geq 1$ and $m \geq 2$ such that $a_k^2 + c^3$ is the mth power of some positive integer.

(Bulgaria)

Solutions

25.1. Since $YM \perp AC$, it is sufficient to prove that $OF \perp XF$. We will prove this by using coordinates. Take the origin of the orthogonal system of coordinates in F so that we have the coordinates: $F(0,0), A(a,0), B(-b,0), C(0,c)$ and $P(-a,0)$, where a, b, c are positive numbers and $a > b$. Assume that $H(0,h)$. From the equality of angles $\angle BCF = \angle HAF$ we obtain $\frac{h}{a} = \frac{b}{c}$, giving that $h = \frac{ab}{c}$. In order to find the coordinates of the circumcenter O we can write the equation of the circumcircle, using for example the determinantal formula and we find:

$$x^2 + y^2 - (a-b)x - \frac{c^2 - ab}{c}y - ab = 0.$$

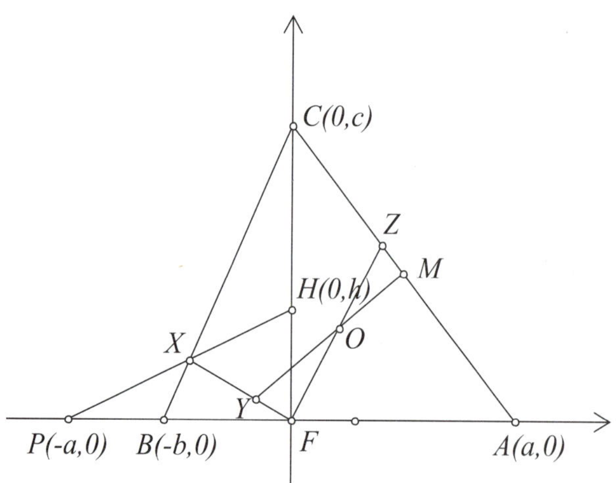

Figure 25.1

It can be rewritten in the form:

$$\left(x - \frac{a-b}{2}\right)^2 + \left(y - \frac{c^2 - ab}{2c}\right)^2 = R^2,$$

showing that O has coordinates $O\left(\frac{a-b}{2}, \frac{c^2-ab}{2c}\right)$.

For the coordinates of X we use Menelaus' theorem in the triangle CBF for the transversal line PH. We have:

$$\frac{BX}{XC} \cdot \frac{CH}{HF} \cdot \frac{FP}{PB} = 1.$$

From this equality one obtains:

$$\lambda = \frac{BX}{XC} = \frac{PB}{FP} \cdot \frac{HF}{CH} = \frac{h(a-b)}{a(c-h)} = \frac{b(a-b)}{c^2 - ab}.$$

Taking in account that F is the origin of the coordinate system we have the coordinates of X in vectorial form:

$$\overrightarrow{FX} = \frac{1}{\lambda+1}\overrightarrow{FB} + \frac{\lambda}{\lambda+1}\overrightarrow{FC} = \left(\frac{b(ab-c^2)}{c^2-b^2}, \frac{bc(a-b)}{c^2-b^2}\right).$$

The vector $\overrightarrow{FO}$ is

$$\overrightarrow{FO} = \left(\frac{a-b}{2}, \frac{c^2-ab}{2c}\right).$$

It is easy to check that $\overrightarrow{FX} \cdot \overrightarrow{FO} = 0$.

Second solution. Actually, since all we have to prove is that $OF \perp FX$, we don't really need points Y and Z. Let X' be the reflection of point X across CF (see Figure 25.2). Since the points O and H are isogonal conjugates with respect to triangle ABC it follows that $\angle CAO = \angle BAH$, hence $\angle CAO = \angle FAX'$. Also, $\angle BCH = \angle ACO$, but since $\angle BCH = \angle X'CH$, we deduce that $\angle FCO = \angle ACX'$, therefore points O and X' are also isogonal conjugates with respect to triangle ACF.

But then

$$\angle HFO = \angle X'FA = \angle XFB,$$

and since $HF \perp BF$, it follows that $OF \perp FX$, as desired.

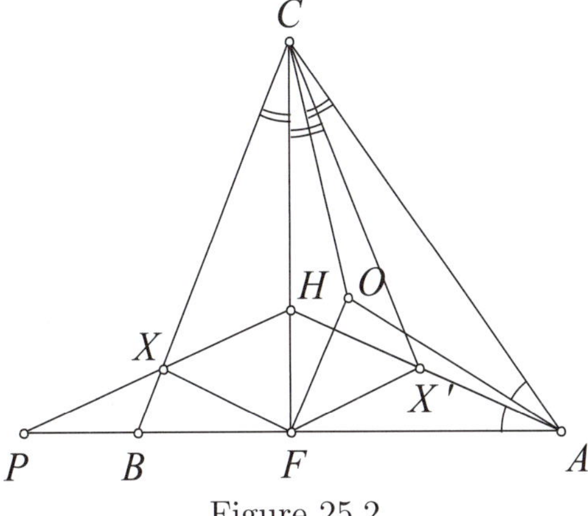

Figure 25.2

25.2. We will prove that such a sequence does not exist.

First solution. We split the sequence in subsets containing the numbers

$$a_{2^i+1}, a_{2^i+2}, \ldots, a_{2^{i+1}},$$

for all $i = 0, 1, 2, \ldots.$ By the HM-AM inequality applied to this set of numbers we have:

$$\sum_{k=2^i+1}^{2^{i+1}} \frac{1}{a_k} \geq \frac{2^{2i}}{\sum\limits_{k=2^i+1}^{2^{i+1}} a_k} > \frac{2^{2i}}{\sum\limits_{k=1}^{2^{i+1}} a_k} \geq \frac{2^{2^i}}{2^{2i+2}} = \frac{1}{4}.$$

Using this inequality for all $i = 0, 1, \ldots, n-1$ we have:

$$\sum_{k=1}^{2^n} \frac{1}{a_k} = \frac{1}{a_1} + \sum_{i=0}^{n-1} \left(\sum_{k=2^i+1}^{2^{i+1}} \frac{1}{a_k} \right) > \frac{1}{a_1} + \frac{n}{4}.$$

This contradicts the condition (ii).

Second solution. Denote $s_n = \sum\limits_{i=1}^{n} \frac{1}{a_i}$. The condition (ii) states that the sequence $(s_n)_{n\geq 1}$ is bounded. Since it is also monotonic increasing it is convergent and then it is a Cauchy sequence. By the definition of a Cauchy sequence, for $\varepsilon = \frac{1}{4}$, there exists a positive integer N such that for all $m > n \geq N$, one has $s_m - s_n < \frac{1}{4}$. By Cauchy-Schwartz inequality we have:

$$\sum_{k=N+1}^{2N} a_k \geq \frac{N^2}{\sum\limits_{k=N+1}^{2N} \frac{1}{a_k}} = \frac{N^2}{s_{2N} - s_N} > 4N^2.$$

On the other hand, by condition (i) we have:

$$\sum_{k=N+1}^{2N} a_k < \sum_{k=1}^{2N} a_k \leq 4N^2.$$

This is a contradiction.

25.3. The points of S will be called grid points. We split the lines passing through two grid points in four categories, showing that each category contains a number of elements which is a multiple of four.

Let O be the center of the rectangle $ABCD$. Consider a system of coordinates such that O is the origin and the axes are parallel to the sides of the rectangle. We assume that the vertices of the rectangle have coordinates: $A(45n+5/2, 45n+1/2), B(45n+5/2, -45n-1/2), C(-45n-5/2, -45n-1/2)$ and $D(-45n-5/2, 45n+1/2)$. Then the grid points P have coordinates $P(k+1/2, l+1/2)$, where k, l are integers, $-45n-2 \le k \le 45n+2$, $-45n \le l \le 45n$. To determine all types of lines we consider the square of side $90n+1$ inscribed in the rectangle such that its vertices have the coordinates:
$A'(45n+1/2, 45n+1/2), B'(45n+1/2, -45n-1/2), C'(-45n-1/2, -45n-1/2)$, $D'(-45n-1/2, 45n+1/2)$.

The lines of first kind are the lines parallel to the sides of the rectangle (Figure 25.3). The number of these lines is:

$$(90n+2)+(90n+6)=180n+8=4(45n+2)$$

The lines of second type are the lines which connect two grid points, are not parallel to the axes and which do not pass through the point O. Taking the reflections of such a line in the center of the rectangle and in the symmetry axes of the rectangle one obtains three more lines, giving an orbit of four lines. So, the total number of lines of second kind is a multiple of four (Figure 25.4).

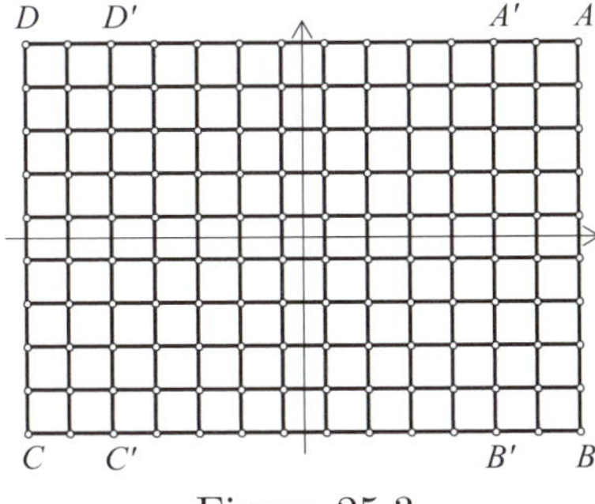

Figure 25.3

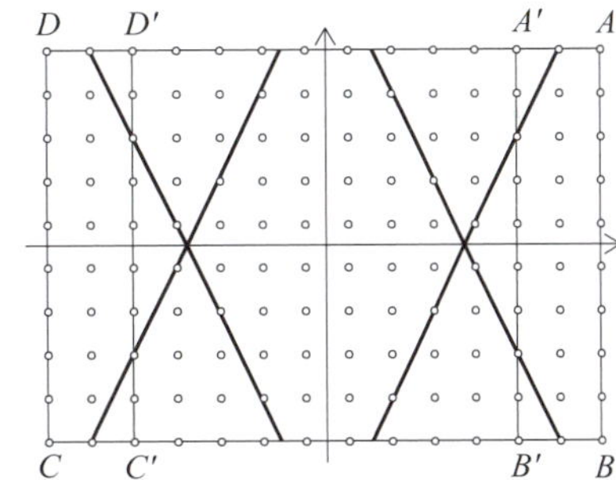

Figure 25.4

The lines of third type are the lines which pass through the center of the rectangle and a grid point of the square, other than the vertices A', B', C', D'. By the symmetry, such a line passes through a second grid point of the square. Taking the reflections of the line in the diagonals of the square and in the axes of symmetry of the rectangle one obtains an orbit containing four distinct lines. So the total number of lines of this type is a multiple of four. It is clear that the diagonals of the square have not been counted because they give an orbit with two lines (Figure 25.5).

The fourth type of lines consists in lines which connect the origin O and a grid point which does not belong to the square. It is clear that these lines pass either through a point of coordinates $P(45n+3/2, l+1/2)$ or $Q(45n+5/2, l+1/2)$ (Figure 25.6).

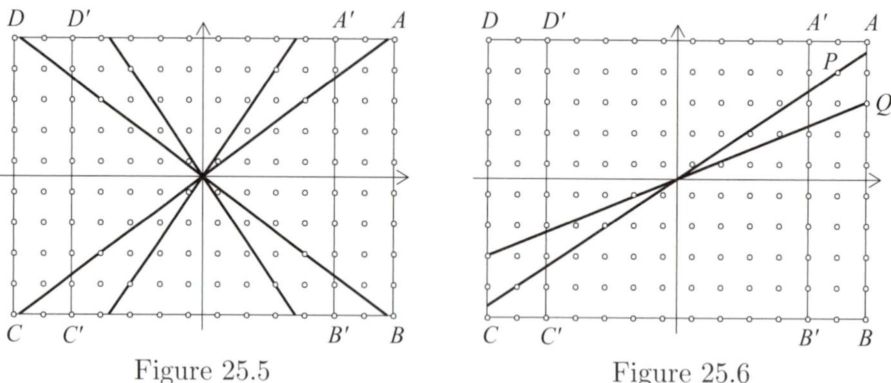

Figure 25.5 Figure 25.6

Such a line is completely characterized by its slope:

$$\lambda = \frac{2l+1}{90n+3} \quad \text{or} \quad \mu = \frac{2l+1}{90n+5}.$$

If a fraction from above is irreducible then the corresponding line does not pass through a point of the square, so it is not a line of third type. Therefore, the number of lines of slope λ is equal to the number of odd numbers $2l+1$ which are relatively prime with $90n+3$. Similarly, we have to compute the number of numbers $2l+1$ which are relatively prime with $90n+5$. We take in account that $-45n \le l \le 45n$, so that $-(90n+1) \le 2l+1 \le 90n+1$. Because

$$\gcd(2l+1, 90n+3) = \gcd(2l+1+90n+3, 90n+3)$$

it follows that we can replace the negative odd numbers $2l+1$, $-(90n+1) \le 2l+1 \le -1$, with the positive even numbers $2k$, $2 \le 2k \le 90n+2$. So, we have to count the number $\varphi(90n+3)$, where φ is Euler's totient function. Similarly, for the lines of slope μ we have to count the number of positive integers q, $1 \le q \le 90n+2$ which are relatively prime to $90n+5$, and this number is $\varphi(90n+5)-2$. Finally, adding the two diagonals of the square, the problem reduces to show that $\varphi(90n+3)+\varphi(90n+5)$ is a multiple of four. We have

$$\varphi(90n+3) = \varphi(3)\varphi(30n+1) = 2\varphi(30n+1),$$

and this is a multiple of four since $\varphi(30n+1)$ is always an even number. Also,

$$\varphi(90n+5) = \varphi(5)\varphi(18n+1).$$

The proof is complete.

25.4. Let k be the least positive integer such that $a_k^2 + c^3$ is a perfect power. We will show that $k = 1$. In this case, $a_1^2 + c^3 = c^2 + c^3 = c^2(c+1)$ is a perfect power if and only if $c + 1$ is a perfect square. This will solve the problem.

Assume by contradiction that $k > 1$. Then $a_k^2 + c^3 = a_{k+1} - a_k$. On the other hand,

$$a_{k+1} - a_k = a_k^2 + a_k - a_{k-1}^2 - a_{k-1} = (a_k - a_{k-1})(a_k + a_{k-1} + 1).$$

Observe that a_{k-1} is defined, as $k > 1$. We will show that $a_k - a_{k-1}$ and $a_k + a_{k-1} + 1$ are relatively prime. Indeed, let p be a common prime factor of these numbers. Then

$$p|(a_k - a_{k-1}) + (a_k + a_{k-1} + 1) = 2a_k + 1 \text{ and}$$

$$p|(a_k + a_{k-1} + 1) - (a_k - a_{k-1}) = 2a_{k-1} + 1.$$

It is easy to obtain the following identity:

$$2(2a_k + 1) = (2a_{k-1} + 1)^2 + 4c^3 + 1.$$

This shows that $p|4c^3 + 1$. On the other hand, the above identity is true for all $k > 1$. Hence, writing it for $k - 1$ we obtain that $p|2a_{k-2} + 1$. Going down, as long as $k - i > 1$ we obtain that $p|2a_1 + 1 = 2c + 1$. Now, since $\gcd(2c + 1, 4c^3 + 1) = 1$, it follows that $p = 1$; this is a contradiction.

Since the numbers $a_k - a_{k-1}$ and $a_k + a_{k-1} + 1$ are relatively prime, it follows from the identity

$$a_k^2 + c^3 = (a_k - a_{k-1})(a_k + a_{k-1} + 1)$$

that $a_k - a_{k-1}$ is a perfect power. But $a_k - a_{k-1} = a_{k-1}^2 + c^3$. This contradicts the selection of k. Therefore, the conclusion is: $a_1^2 + c^3$ is a perfect power if and only if $c + 1$ is a perfect square. We can prove more: a_1 is the unique member of the sequence with this property.

Indeed, for $k > 1$ we have $a_k^2 < a_k^2 + c^3 < (a_k + 1)^2$.

The 26th BMO

The twenty-sixth Balkan Mathematical Olympiad for high-school students was held between April 29th and May 4th, 2007, in the city of Kragujevac, Serbia. The number of the participating countries was 18: 11 member countries (Albania, Bosnia & Herzegovina, Bulgaria, Cyprus, Former Yugoslav Republic of Macedonia, Greece, Republic of Moldova, Montenegro, Romania, Serbia and Turkey) and 7 invited countries (Azerbaijan, France, Italy, Kazakhstan, Tajikistan, Turkmenistan and United Kingdom). A second team from Serbia as well as a team from the city of Brno (Czech Republic) also attended the competition.

Problems

26.1. Solve the equation

$$3^x - 5^y = z^2$$

in positive integers.

(Greece)

26.2. Let MN be a line parallel to the side BC of triangle ABC, with M on the side AB and N on the side AC. The lines BN and CM meet at point P. The circumcircles of triangles BMP and CNP meet at two distinct points P and Q. Prove that $\angle BAQ = \angle CAP$.

(Moldova)

26.3. A rectangle 9×12 is partitioned into unit squares, and the centers of all these squares, except for the four corner squares and the eight squares orthogonally adjacent to them, are colored in red. Is it possible to label these red centers $C_1, C_2, \ldots, C_{96}$ in such a way that the following two conditions are both fulfilled

(i) the distances C_1C_2, $C_2C_3, \ldots,$ $C_{95}C_{96}$, $C_{96}C_1$ are all equal to $\sqrt{13}$,

(ii) the closed broken line $C_1C_2 \ldots C_{96}C_1$ has a center of symmetry?

(Bulgaria)

26.4. Denote by S the set all positive integers.

Find all functions $f : S \to S$ such that

$$f\left(f(m)^2 + 2f(n)^2\right) = m^2 + 2n^2,$$

for all m, n in S.

(Bulgaria)

Solutions

26.1. If x is odd, then we obtain

$$z^2 \equiv 2 \pmod 4,$$

a contradiction. Hence x must be even. Let $x = 2a$; we obtain

$$(3^a - z)(3^a + z) = 5^y,$$

therefore

$$3^a - z = 5^b,\ 3^a + z = 5^c,$$

for some nonnegative integers $b \le c$. But then

$$2 \cdot 3^a = 5^b + 5^c,$$

which cannot hold unless $b = 0$.

Checking the equation

$$2 \cdot 3^a = 5^c + 1$$

modulo 3, we deduce that c must be odd. Moreover, if $a > 1$, the same equation modulo 9 shows that c is a multiple of 3, as well. But then

$$5^{3(2k+1)} + 1 \equiv (-1)^{2k+1} + 1 \equiv 0 \pmod 7,$$

hence 7 divides $2 \cdot 3^a$, which is impossible. We deduce that $a = 1$, hence $c = 1$, so $x = 2$, $y = 1$, $z = 2$.

26.2. We first prove that the quadrilateral $ABQN$ is cyclic. Indeed, since $BMPQ$ and $CNPQ$ are cyclic, we have

$$\angle BQN = \angle BQP + \angle PQN = \angle AMC + \angle ACM = 180^\circ - \angle A,$$

hence $ABQN$ is cyclic. Similarly, one can show that $ACQM$ is cyclic, as well.

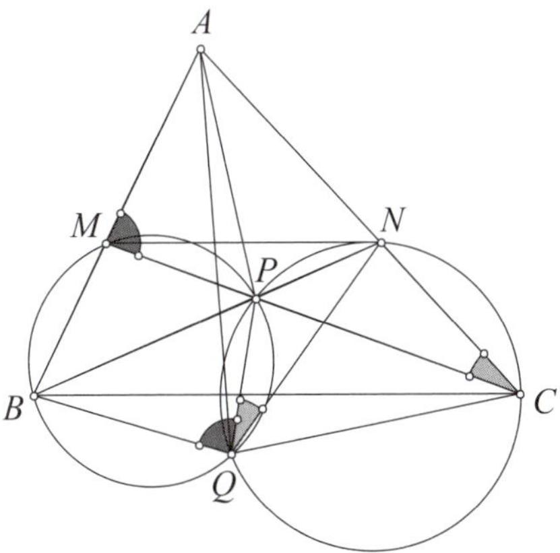

Figure 26.1

(In fact, as the geometry trained reader knows, the point Q is the *Miquel point* of the complete quadrilateral $AMPNBC$, a point through which all circumcircles of the triangles ABN, ACM, BPM, and CPN pass.)

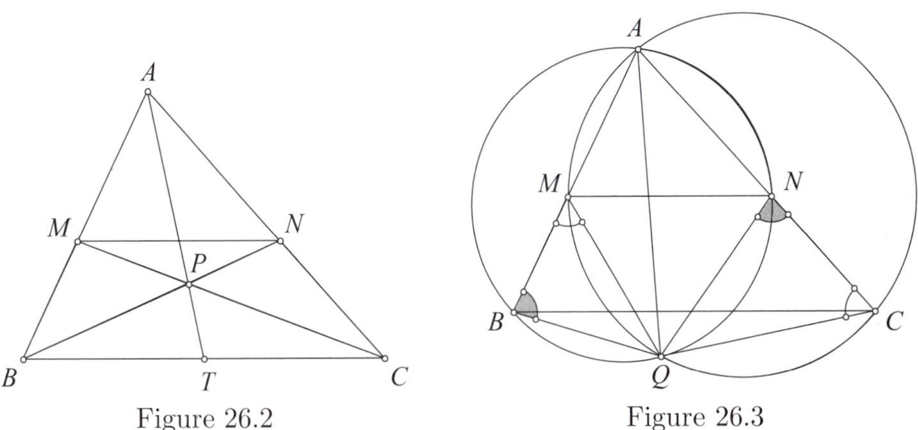

Figure 26.2

Figure 26.3

Let T be the point where the line AP meets BC (see Figure 26.2). Then T is the midpoint of the line segment BC (this follows either from Ceva's theorem in triangle ABC, or from the well-known fact that in a trapezoid the midpoints of the basis, the point of intersection of the diagonals and the point of intersection of the non-parallel sides are collinear).

We deduce that $[ABT] = [ACT]$, hence

$$AB \cdot AT \cdot \sin \angle BAP = AC \cdot AT \cdot \sin \angle CAP,$$

or

$$\frac{\sin \angle BAP}{\sin \angle CAP} = \frac{AC}{AB}.$$

Applying the sine law in triangles ABQ and ANQ (note that they have the same circumcircle), we obtain

$$\frac{\sin \angle BAQ}{BQ} = \frac{\sin \angle CAQ}{NQ},$$

hence

$$\frac{\sin \angle BAQ}{\sin \angle CAQ} = \frac{BQ}{NQ}.$$

But the triangles BMQ and NCQ are similar (this follows from the fact that $AMQC$ and $ANQB$ are cyclic quads; see Figure 26.3), therefore

$$\frac{BQ}{NQ} = \frac{BM}{NC} = \frac{AB}{AC},$$

and we conclude that

$$\frac{\sin \angle BAQ}{\sin \angle CAQ} = \frac{AB}{AC}.$$

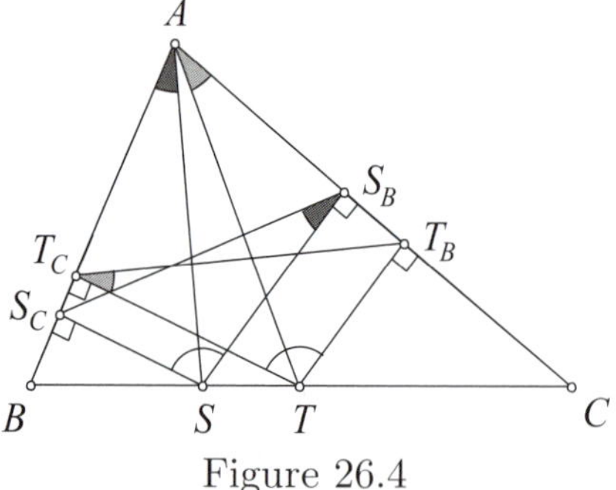

Figure 26.4

Finally, denote by S the intersection between AQ and BC. Let us focus on the triangle ABC, with the points S and T on BC. Project S and T on AC and BC (see Figure 26.4). We have

$$\frac{AB}{AC} = \frac{\sin \angle BAQ}{\sin \angle CAQ} = \frac{\sin \angle BAS}{\sin \angle CAC} = \frac{SS_C}{SS_B},$$

and similarly,

$$\frac{AC}{AB} = \frac{TT_C}{TT_B}.$$

Note that the quadrilaterals AS_BSS_C and AT_BTT_C are cyclic, hence $\angle S_BSS_C = \angle T_BTT_C = 180^\circ - \angle A$. Since we also have

$$\frac{TT_B}{TT_C} = \frac{SS_C}{SS_B},$$

it follows that the triangles SS_BS_C and TT_CT_B are similar, hence $\angle S_CS_BS = \angle T_BT_CT$. But, in the cyclic quadrilaterals AS_BSS_C and AT_BTT_C, we have $\angle S_CS_BS = \angle BAS$, and $\angle T_BT_CT = \angle CAT$.

We deduce that $\angle BAS = \angle CAT$, that is, $\angle BAQ = \angle CAP$, as desired.

Second solution. We will use complex numbers. Let us denote by x the affix of the point X in the complex plane.

In terms of complex numbers, we have to prove that

$$\arg\frac{q-a}{b-a} = \arg\frac{c-a}{p-a} = \arg\frac{c-a}{t-a},$$

which is equivalent to

$$\frac{q-a}{b-a} : \frac{c-a}{t-a} \in \mathbb{R}_+.$$

First, choose the origin of the complex plane in A, so that $a = 0$. Since MN is parallel to BC, there exists $\lambda > 0$ such that $m = \lambda b$ and $n = \lambda c$. As we saw in the previous solution, triangles BMQ and NCQ are similar, hence

$$\frac{m-b}{q-b} = \frac{c-n}{q-n},$$

yielding

$$q = \frac{nm-bc}{m+n-b-c} = \frac{bc\left(\lambda^2-1\right)}{(b+c)\left(\lambda-1\right)} = \frac{bc\left(\lambda+1\right)}{b+c}.$$

Since T is the midpoint of BC, we have $t = \frac{b+c}{2}$, and we have

$$\frac{q-a}{b-a} : \frac{c-a}{t-a} = \frac{q}{b}\cdot\frac{t}{c} = \frac{\frac{bc(\lambda+1)}{b+c}}{b}\cdot\frac{\frac{b+c}{2}}{c} = \frac{\lambda+1}{2} \in \mathbb{R}_+.$$

26.3. Actually, the problem asks about the existence of a central symmetric hamiltonian circuit in the graph whose vertices are the 96 points in our grid and in which two vertices are connected by an edge if and only if the distance between them is $\sqrt{13}$. We will show that no such circuit exists. Suppose the contrary.

First, color the red point squares in a black and white checkerboard pattern, and observe that two red points $\sqrt{13}$ apart lie in squares of different colors, hence black and white alternate along the circuit. Consider the points

$A(2,2)$ and $B(8,11)$ (see Figure 26.5). The circuit can be divided into two parts: one leading from A to B, the other from B to A. If the two parts are symmetric to each other, each must consist of $96/2 = 48$ edges. But if an even number of edges connect A and B, then the two points must lie in squares of the same color, which is obviously not the case. We conclude that each of the two parts consists of an odd number of edges and, moreover, each is symmetric to itself. We deduce that each of the two parts must contain at least one self-symmetric edge. The only edges with this property are those joining $(4,5)$ to $(6,8)$, and $(6,5)$ to $(4,8)$, therefore these two must belong to the circuit.

For each point in the grid exactly two edges in our circuit emerge from it. Consider now the point $A(2,2)$. It can be connected only to $(5,4)$ and $(4,5)$, so both these two edges must belong to the circuit.

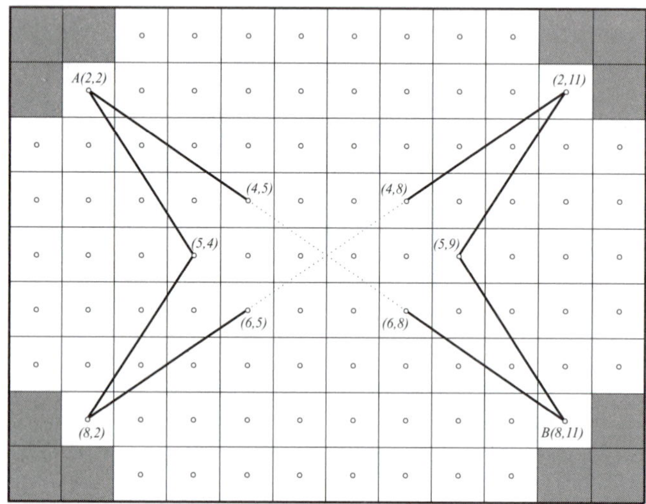

Figure 26.5

Similar considerations for the points $(8,2)$, $(8,11)$, and $(2,11)$ show that the circuit must include the edges $(2,2)-(5,4)-(8,2)-(6,5)-(4,8)-(2,11)-(5,9)-(8,11)-(6,8)-(4,5)-(2,2)$.

But this is a closed circuit which does not contain all the 96 points, which is a contradiction.

Observation. The existence of a non-central symmetric hamiltonian circuit has been settled by a computer-found result[6] (see Figure 26.6)

[6] Thanks to Dan Schwarz and Codruţ Grosu.

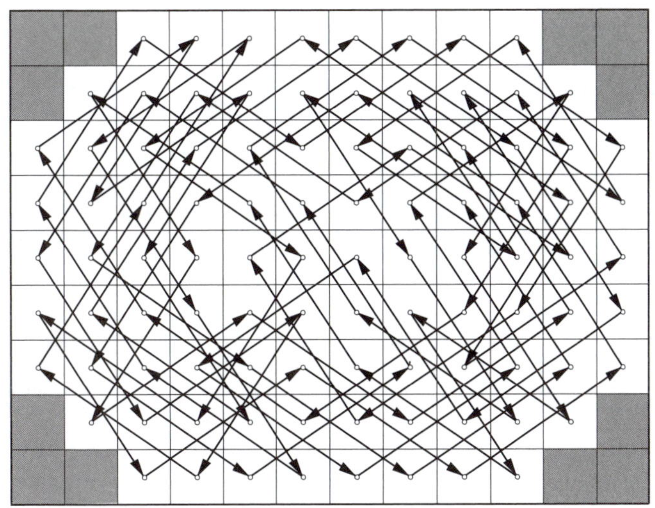

Figure 26.6

26.4. First, observe that f is a one-to-one function. Indeed, if m_1, m_2 are positive integers and $f(m_1) = f(m_2)$, then for any positive integer n, we have $f\left(f(m_1)^2 + 2f(n)^2\right) = f\left(f(m_2)^2 + f(n)^2\right)$, hence $m_1^2 + n^2 = m_2^2 + n^2$, that is, $m_1 = m_2$.

This implies the equivalence

$$f(m)^2 + 2f(n)^2 = f(p)^2 + 2f(q)^2 \Longleftrightarrow m^2 + 2n^2 = p^2 + 2q^2 \qquad (*)$$

Let $f(1) = a$ and plug $m = n = 1$ in the original equation. This yields

$$f\left(3a^2\right) = 3.$$

Now, observe that

$$\left(5a^2\right)^2 + 2\left(a^2\right)^2 = \left(3a^2\right)^2 + 2\left(3a^2\right)^2,$$

hence, by (*), we have

$$f\left(5a^2\right)^2 + 2f\left(a^2\right)^2 = f\left(3a^2\right)^2 + 2f\left(3a^2\right)^2 = 3f\left(3a^2\right)^2 = 3 \cdot 9 = 27.$$

As one can easily check, the solutions in positive integers of the diophantine equation $x^2 + 2y^2 = 27$ are $(3,3)$ and $(5,1)$. We cannot have $f\left(5a^2\right) = f\left(a^2\right) = 3$, since f is one-to-one, hence $f\left(5a^2\right) = 5$ and $f\left(a^2\right) = 1$.

Again by (*) we have

$$f\left(a^2\right)^2 + 2f\left(4a^2\right)^2 = f\left(5a^2\right)^2 + 2f\left(2a^2\right)^2,$$

whence

$$2f\left(4a^2\right)-2f\left(2a^2\right)=24.$$

The only solution in positive integers of the equation $x^2-y^2=12$ is $(4,2)$, therefore $f\left(4a^2\right)=4$ and $f\left(2a^2\right)=2$.

We will prove by induction that $f\left(ka^2\right)=k$, for all positive integers k. As we have seen, this holds for $k\in\{1,2,3,4,5\}$. For the inductive step, observe that for all k we have

$$(k+4)^2+2(k+1)^2=k^2+2(k+3)^2,$$

therefore

$$f\left((k+4)a^2\right)^2=2f\left((k+3)a^2\right)^2-2f\left((k+1)a^2\right)^2+f\left(ka^2\right)^2,$$

from which we deduce our claim, since

$$k+4=2(k+3)-2(k+1)+k.$$

Finally, since $f\left(ka^2\right)=k$, for all positive integers k, we also have $f\left(a^3\right)=a=f(1)$, hence $a=1$, and then $f(k)=k$ for all k in S. Clearly, this function satisfies the given condition.

Second solution. Again, we start by noticing that f is one-to-one and therefore (*) holds. Observe that

$$(n+4)^2+2(n+1)^2=n^2+2(n+3)^2,$$

for all positive integers n. Therefore, setting $x_n=f(n)^2$, we obtain

$$x_{n+4}=2x_{n+3}-2x_{n+1}+x_n,$$

for all n.

The characteristic equation of this recursive relation is

$$(x+1)(x-1)^3=0,$$

hence

$$x_n=an^2+bn+c+d(-1)^n,$$

for some constants a,b,c, and d. Let $y_n=\sqrt{x_n}$. Since x_n is the square of a positive integer, it follows that (y_n) is a sequence of positive integers.

Observe that

$$\begin{aligned} y_{n+1} - y_n = \sqrt{x_{n+1}} - \sqrt{x_n} &= \frac{x_{n+1} - x_n}{\sqrt{x_{n+1}} + \sqrt{x_n}} \\ &= \frac{2an + a + b + 2d\,(-1)^n}{\sqrt{a\,(n+1)^2 + b\,(n+1) + \ldots} + \sqrt{an^2 + bn + \ldots}} \\ &= \frac{2an + a + b + 2d\,(-1)^n}{n\sqrt{a}\left(\sqrt{\left(1 + \frac{1}{n}\right)^2 + b\left(\frac{1}{n} + \frac{1}{n^2}\right) + \ldots} + \sqrt{1 + \frac{b}{n} + \ldots}\right)}, \end{aligned}$$

hence

$$\lim_{n \to \infty} (y_{n+1} - y_n) = \sqrt{a}.$$

Since (y_n) is a convergent sequence of positive integers, its limit must be a positive integer as well, and the sequence must eventually become constant. We deduce that $a = b^2$, for some positive integer b, and that $f(n) = bn$, for all n. Plugging in the original equation yields $b = 1$, hence $f(n) = n$, for all n in S.

The 27th BMO

The twenty-seventh Balkan Mathematical Olympiad for high-school students was held between May 2nd and May 8th, 2010, in Chişinău, Republic of Moldova. The number of the participating countries was 18: 10 member countries (Albania, Bulgaria, Cyprus, Former Yugoslav Republic of Macedonia, Greece, Republic of Moldova, Montenegro, Romania, Serbia and Turkey) and 8 invited countries (Azerbaijan, France, Italy, Kazakhstan, Saudi Arabia, Tajikistan, Turkmenistan and United Kingdom). A second team from Moldova also attended the competition.

Problems

27.1. Let a, b and c be positive real numbers. Prove that

$$\frac{a^2b\,(b-c)}{a+b}+\frac{b^2c\,(c-a)}{b+c}+\frac{c^2a\,(a-b)}{c+a}\geq 0.$$

(Saudi Arabia)

27.2. Let ABC be an acute triangle with orthocenter H, and let M be the midpoint of AC.

The point C_1 on AB is such that CC_1 is an altitude of the triangle ABC. Let H_1 be the reflection of H in AB. The orthogonal projections of C_1 onto the lines AH_1, AC and BC are P, Q and R, respectively. Let M_1 be the point such that the circumcentre of triangle PQR is the midpoint of the segment MM_1.

Prove that M_1 lies on the segment BH_1.

(Serbia)

27.3. A *strip* of width w is the set of all points which lie on, or between, two parallel lines distance w apart. Let S be a set of n $(n \geq 3)$ points on the plane such that any three different points of S can be covered by a strip of width 1.

Prove that S can be covered by a strip of width 2.

(Romania)

27.4. For each integer n $(n \geq 2)$, let $f(n)$ denote the sum of all positive integers that are at most n and not relatively prime to n.

Prove that $f(n+p) \neq f(n)$ for each such n and every prime p.

(Turkey)

Solutions

27.1. After dividing by abc, we obtain the equivalent inequality

$$\frac{a(b-c)}{c(a+b)} + \frac{b(c-a)}{a(b+c)} + \frac{c(a-b)}{b(c+a)} \geq 0.$$

Observe that

$$\frac{a(b-c)}{c(a+b)} + 1 = \frac{b(a+c)}{c(a+b)},$$

so, adding 3 to both sides yields

$$\frac{b(a+c)}{c(a+b)} + \frac{c(a+b)}{a(b+c)} + \frac{a(b+c)}{b(c+a)} \geq 3,$$

which is an immediate consequence of the $AM - GM$ inequality:

$$\frac{b(a+c)}{c(a+b)} + \frac{c(a+b)}{a(b+c)} + \frac{a(b+c)}{b(c+a)} \geq 3\sqrt[3]{\frac{b(a+c)}{c(a+b)} \cdot \frac{c(a+b)}{a(b+c)} \cdot \frac{a(b+c)}{b(c+a)}} = 3.$$

Second solution. Clearing denominators leads to the following equivalent inequality:

$$a^3b^3 + b^3c^3 + c^3a^3 \geq a^2bc^3 + b^2ca^3 + c^2ab^3,$$

which can be justified using again $AM - GM$. Indeed, by $AM - GM$, we have

$$a^3c^3 + a^3c^3 + b^3c^3 \geq 3a^2bc^3,$$
$$a^3b^3 + a^3b^3 + a^3c^3 \geq 3b^2ca^3,$$
$$a^3b^3 + b^3c^3 + b^3c^3 \geq 3c^2ab^3.$$

Adding up these three inequalities gives the desired result.

Third solution. Another approach is based on the rearrangements inequality. We can assume, with no loss of generality, that $a = \max(a, b, c)$, and analyze two cases: $a \geq b \geq c$, and $a \geq c \geq b$.

Dividing again by abc, we can write the inequality in the equivalent form

$$\frac{ab}{c(a+b)} + \frac{bc}{a(b+c)} + \frac{ca}{b(c+a)} \geq \frac{a}{a+b} + \frac{b}{b+c} + \frac{c}{c+a}.$$

If $a \geq b \geq c$, we apply the rearrangements inequality to the ordered triples (ab, ac, bc) and $\left(\frac{1}{c(a+b)}, \frac{1}{a(b+c)}, \frac{1}{b(c+a)}\right)$.

If $a \geq c \geq b$, the triples are (ac, ab, bc) and $\left(\frac{1}{b(c+a)}, \frac{1}{c(a+b)}, \frac{1}{a(b+c)}\right)$, and we are done.

We can make things a bit simpler by replacing a, b, c with $\frac{1}{x}, \frac{1}{y}, \frac{1}{z}$. This yields

$$\frac{1}{xyz}\left(\frac{z-y}{x+y} + \frac{x-z}{y+z} + \frac{y-x}{z+x}\right) \geq 0,$$

or, equivalently

$$\frac{x}{y+z} + \frac{y}{z+x} + \frac{z}{x+y} \geq \frac{z}{y+z} + \frac{x}{z+x} + \frac{y}{x+y},$$

which again follows from the rearrangements inequality.

27.2. If K is a point inside a triangle ABC, and K_A, K_B, K_C are its projections onto the triangle's sides, then the triangle $K_AK_BK_C$ is called the *pedal triangle* of the point K (with respect to the triangle ABC). We will use a classical result.

Lemma. Let K be a point in the interior of the triangle ABC and let L be its isogonal conjugate. Then the pedal triangles of points K and L have the same circumcircle, whose center is the midpoint of the line segment KL.

Proof. With the notations in Figure 27.1, let us show that the quadrilateral $K_CL_CK_BL_B$ is cyclic.

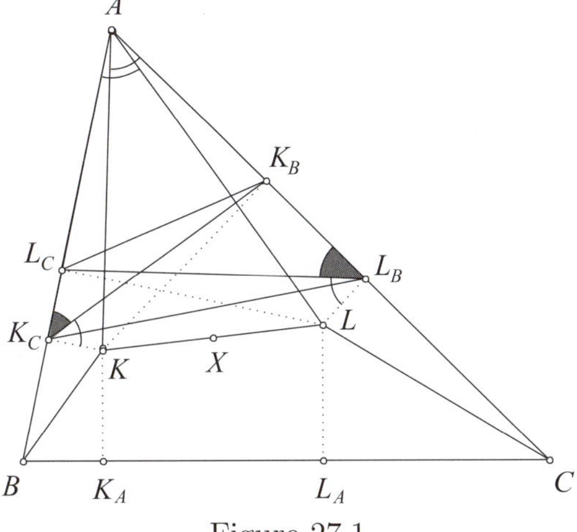

Figure 27.1

Since the angles $\angle KK_CA$ and $\angle KK_BA$ are right, KK_CAK_B is a cyclic quad, hence $\angle KK_CK_B = \angle KAK_B$. Similarly, we have $\angle LL_BL_C = \angle LAL_C$. But K and L are isogonal conjugates, hence $\angle KAK_B = \angle LAL_C$. It follows that $\angle KK_CK_B = \angle LL_BL_C$, that is, $K_CL_CK_BL_B$ is cyclic.

The center of the circle passing through the points above lies on the perpendicular bisectors of both line segments K_CL_C and K_BL_B, which obviously pass through the midpoint X of the line segment KL. It follows that $XK_B = XK_C$. Similarly,we obtain $XK_A = XK_B$, hence $XK_A = XK_B = XK_C$, and, of course, also $XL_A = XL_B = XL_C$. Since $XK_C = XL_C$, it follows that the six vertices of the two pedal triangles are on a circle centered at X, the midpoint of KL, and this ends the proof of the lemma.

Returning to the problem, let D be the point of intersection between the lines BC and AH_1.

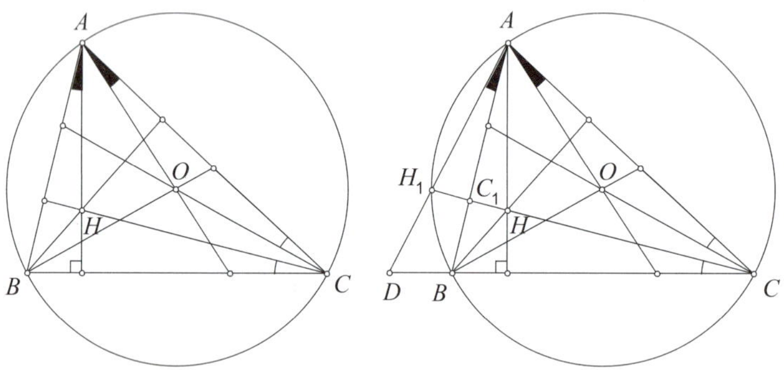

Figure 27.2

Let us observe that in the triangle ABC, the isogonal conjugate of the orthocenter H is the circumcenter O (see Figure 27.2). Using this, it is not difficult to see that the isogonal conjugate of C_1 (in the triangle ACD) is also the circumcenter O of triangle ABC.

From the lemma we deduce that the points P, Q, R, and M lie on a circle centered at X, the midpoint of C_1O. But X is also the midpoint of MM_1, hence the quadrilateral OMC_1M_1 is a parallelogram. It follows that $C_1M_1 = OM$ and $C_1M_1 || OM$.

On the other hand, it is known that $OM || BH$ and

$$OM = \frac{1}{2}BH,$$

(see Appendix, The Euler Line). We deduce that C_1M_1 is a midline in the triangle BHH_1, and therefore M_1 lies on BH_1.

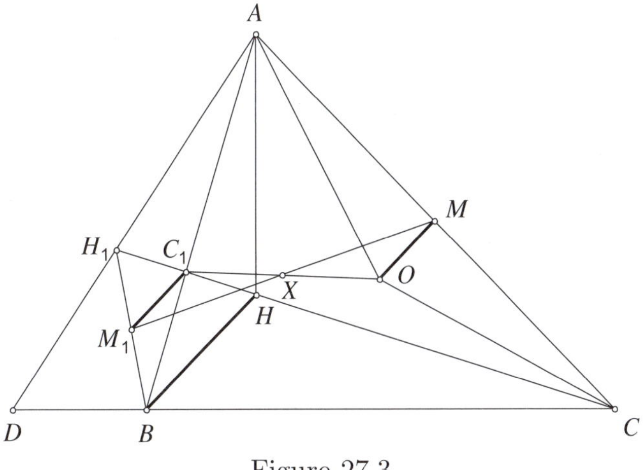

Figure 27.3

27.3. We will use the following
Lemma. A triangle can be covered by a strip of width w if and only if at least one of its altitudes has the length at most w.
Proof. If an altitude of the triangle has the length less than or equal to w, it can be covered with a strip of width w in a trivial way.

Conversely, suppose that the triangle ABC is covered with a strip of width w, and draw through its vertices perpendiculars to the boundaries of the strip. We distinguish two cases: one of these perpendiculars intersects one of the triangle's sides in its interior (see Figure 27.4) or one of the triangle's sides is contained in one of the perpendiculars (Figure 27.5). It is not difficult to see that in both cases one of the triangle's altitudes (in our figures, AD) has the length at most w.

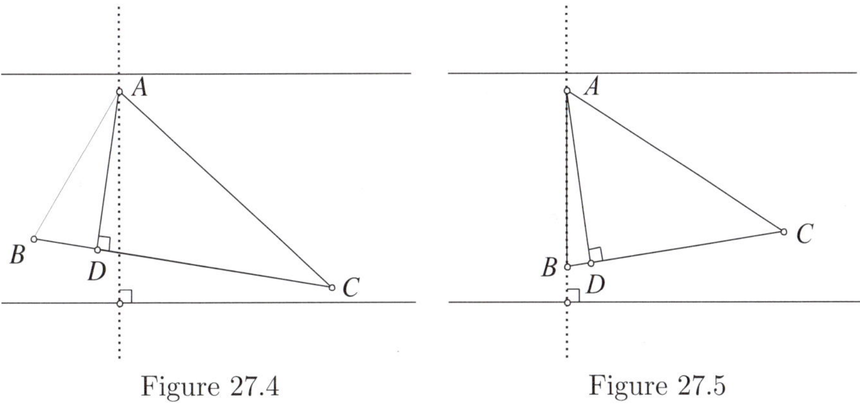

Figure 27.4

Figure 27.5

Returning to the problem, let A and B be two points in S such that the distance AB is maximal. If C is another point in S, then the distance d from C to the line AB is at most 1. Indeed, if C does not lie on AB, then d is the shortest altitude of the triangle ABC, since AB is the longest side. According to the lemma, $d \leq 1$. Obviously, if C lies on AB, then $d = 0$.

It follows that all points in S are covered by a strip of width 2, formed by the two parallels to AB, each at distance 1 from AB.

Another idea is to consider the points A, B, and C in S such that the area of the triangle ABC is maximal. Draw through each vertex of the triangle a parallel to the opposite side, obtaining thus the triangle $A'B'C'$ (see Figure 27.6).

We claim that $A'B'C'$ covers all points in S. Indeed, if M is a point in S lying in the exterior of the triangle $A'B'C'$, then at least one of the areas of triangles ABM, BCM, and ACM is greater than the area of ABC, contradicting thus the choice of ABC (in Figure 27.6 we have $[ABM] > [ABC]$).

Finally, since one of the altitudes of ABC has the length at most 1, it follows that one of the altitudes of $A'B'C'$ has the length at most 2, and we are done.

Observations. 1. Using the previous construction, we can obtain the conclusion without the stated lemma.

Indeed, the line segments AA', BB', and CC' are concurrent at G, the centroid of the triangle ABC. A homothety centered at G, with ratio -2, sends ABC to $A'B'C'$ and a strip of width 1 covering ABC to a strip of width 2 covering $A'B'C'$, and thus, all points in S.

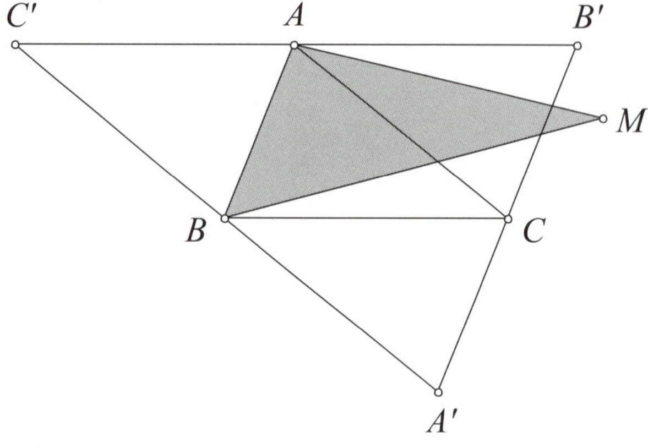

Figure 27.6

2. It seems that the bound 2 for the width of the covering strip can be improved. We don't know a proof for this, but numerical experiments show

that a strip of width $2\cos\frac{\pi}{5}$ still covers S[7]. However, if S contains only four points, one can prove that it suffices a strip of width $\sqrt{2}$ to cover S[8].

27.4. It is not difficult to see that if $1 \leq k \leq n$ and $\gcd(k,n) = 1$, then $\gcd(n-k,n) = 1$. Therefore, if we define

$$A = \{k \in \mathbb{N} |\ k \leq n,\ \ \gcd(k,n) = 1\}$$

then

$$f(n) = 1 + 2 + \ldots + n - \sum_{k \in A} k = \frac{n(n+1)}{2} - \frac{1}{2}\left(\sum_{k \in A} k + \sum_{k \in A}(n-k)\right)$$
$$= \frac{n(n+1)}{2} - \frac{1}{2}n \cdot |A| = \frac{n}{2}(n + 1 - \varphi(n)),$$

where φ denotes Euler's totient function.

We see that the equality $f(n+p) = f(n)$ is equivalent to

$$(n+p)(n+p+1-\varphi(n+p)) = n(n+1-\varphi(n)). \tag{1}$$

If p does not divide n, then $\gcd(n+p,n) = 1$, and hence $n+p$ must divide $n+1-\varphi(n)$, which is not possible, since $1 \leq n+1-\varphi(n) \leq n$. Therefore $n = pm$, for some positive integer m, and (1) becomes

$$(m+1)(mp+p+1-\varphi(mp+p)) = m(mp+1-\varphi(mp)). \tag{2}$$

Clearly, $\gcd(m+1,m) = 1$, hence there exists an integer a such that

$$mp + p + 1 - \varphi(mp+p) = am$$

and

$$mp + 1 - \varphi(mp) = a(m+1).$$

These equalities imply $a \equiv -1 \pmod{p-1}$ and since

$$1 \leq mp + 1 - \varphi(mp) \leq mp,$$

it follows that $1 \leq a \leq p-1$, hence $a = p-2$.

We deduce that

$$2m - \varphi(mp) = p - 3 \tag{3}$$

and

$$2m + p + 1 - \varphi(mp+p) = 0. \tag{4}$$

[7]Thanks to Ciprian Pop for this information.

[8]See American Mathematical Monthly, problem 11247 [2006,760]

We claim that p is relatively prime to both m and $m+1$. Indeed, let $m = p^a s$, with $\gcd(p, s) = 1$; then (3) leads to

$$p^a (2s - (p-1)\varphi(s)) = p - 3,$$

and for $a \geq 1$ this implies that p divides 3, hence $p = 3$ and $s = 1$. But then $m = 3^a$ and from (4) it follows that $\varphi(3^a + 1) = 3^a + 2$, a contradiction.

If p divides $m+1$, then, from (4) again, we deduce that p divides 1, a contradiction.

Thus, $\varphi(mp) = (p-1)\varphi(m)$ and $\varphi(p(m+1)) = (p-1)\varphi(m+1)$, transforming (3) and (4) into

$$\varphi(m) = \frac{2(m+1)}{p-1} - 1, \tag{5}$$

$$\varphi(m+1) = \frac{2(m+1)}{p-1} + 1. \tag{6}$$

Because $\varphi(m+1) < m+1$ it follows that $p \geq 5$. Moreover, if $p = 5$, then $\varphi(m) = \frac{1}{2}(m-1)$, hence m is odd, and $\varphi(m+1) \leq \frac{1}{2}(m+1)$, contradicting (6). We deduce that $p \geq 7$. Also, since $\varphi(m) = 1$ cannot hold, we have

$$m + 1 \geq \frac{1}{2}(p-1)(\varphi(m) + 1) \geq 9.$$

Since $\varphi(m+1) - \varphi(m) = 2$, at most one of the numbers $\varphi(m)$ and $\varphi(m+1)$ is divisible by 4.

If 4 does not divide $\varphi(m)$, then m has at most one odd prime factor, hence $m = 2^u q^v$, for some integers u, v and odd prime q. Since $v \neq 0$ (otherwise $u \geq 3$) we have $u \leq 1$ and, using that $p \geq 7$, we obtain

$$\frac{m+1}{3} - 1 \geq \frac{2(m+1)}{p-1} - 1 = \varphi(m) = (q-1)q^{v-1} \geq \frac{1}{2}\left(1 - \frac{1}{q}\right) \geq \frac{m}{3},$$

a contradiction.

If 4 does not divide $\varphi(m+1)$ we reach a contradiction in a similar way.

The 28th BMO

The twenty-eighth Balkan Mathematical Olympiad for high-school students was held between May 4th and May 8th , 2011, in the city of Iasi, Romania. The number of participating countries was 20: 11 member countries (Albania, Bosnia and Herzegovina, Bulgaria, Cyprus, Former Yugoslav Republic of Macedonia, Greece, Republic of Moldova, Republic of Montenegro, Romania, Serbia and Turkey) and 9 invited countries (Azerbaijan,France, Indonesia, Italy, Kazakhstan, Saudi Arabia, Tajikistan, Turkmenistan and United Kingdom.) Additionally, a second Romanian team took part as Romania 2.

Problems

28.1. Let $ABCD$ be a cyclic quadrilateral which is not a trapezoid and whose diagonals meet at E. The midpoints of AB and CD are F and G, respectively and ℓ is the line through G parallel to AB. The feet of the perpendiculars from E to the lines ℓ and CD are H and K, respectively. Prove that the lines EF and HK are perpendicular.

(United Kingdom)

28.2. Given real numbers x, y, z such that $x + y + z = 0$, show that

$$\frac{x(x+2)}{2x^2+1} + \frac{y(y+2)}{2y^2+1} + \frac{z(z+2)}{2z^2+1} \geq 0.$$

When does the equality hold?

(Greece)

28.3. Let S be a finite set of positive integers which has the following property: if x is a member of S then so are all its positive divisors. A non-empty subset T of S is *good* if whenever $x, y \in T$ and $x < y$, the ratio y/x is a power of a prime number. A non-empty subset T of S is *bad* if whenever $x, y \in T$ and $x < y$, the ratio y/x is not a power of a prime number. We agree that a singleton of S is both good and bad. Let k be the largest possible size of a good subset of S. Prove that k is also the smallest number of pairwise disjoint bad subsets whose union is S.

(Bulgaria)

28.4. Let $ABCDEF$ be a convex hexagon of area 1, whose opposite sides are parallel. The lines AB, CD and EF meet in pairs to determine the vertices of a triangle. Similarly , the lines BC, DE and FA meet in pairs to determine the vertices of another triangle. Show that at least one of these two triangles has area at least $3/2$.

(Bulgaria)

Solutions

28.1. The triangles BEA and CED are similar, because $ABCD$ is a cyclic quadrilateral. The line segments EF and EG are medians in these triangles, respectively. Then the triangles BEF and CEG are also similar. It follows that $\angle BFE = \angle CGE$ (Figure 28.1).

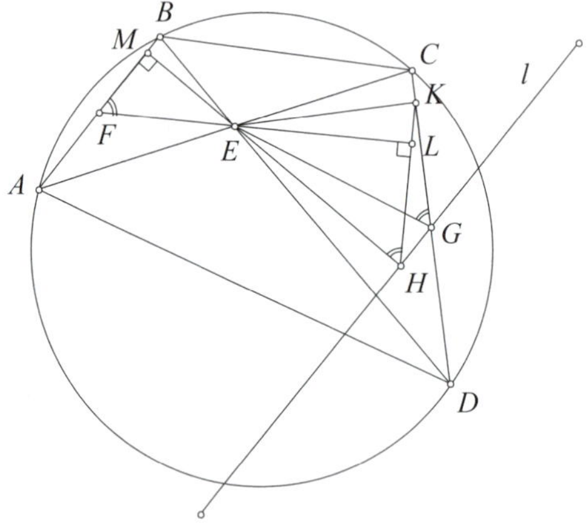

Figure 28.1

The lines CD and ℓ are distinct, so the points H and K are distinct as well. Since the points E, K, H, G are on on the circle of diameter GE it follows that $\angle EGK = \angle EHK$. Hence, we have the equal angles $\angle EHK = \angle EFB$.

Let L be the intersection point of the lines HK and EF and M the intersection point of HE and AB. By the equality of angles $\angle EHK$ and $\angle EFB$, it follows that $FHLM$ is a cyclic quadrilateral. Therefore, $\angle FLH = \angle FMH = 90°$.

28.2. The inequality is obviously true when $xyz = 0$ and in this case the equality holds if and only if $x = y = z = 0$. So, we can assume that $xyz \neq 0$. The inequality is equivalent to

$$\left(\frac{x(x+2)}{2x^2+1}+\frac{1}{2}\right)+\left(\frac{y(y+2)}{2y^2+1}+\frac{1}{2}\right)+\left(\frac{z(z+2)}{2z^2+1}+\frac{1}{2}\right)\geq\frac{3}{2},$$

which rewrites as

$$\frac{(2x+1)^2}{2x^2+1}+\frac{(2y+1)^2}{2y^2+1}+\frac{(2z+1)^2}{2z^2+1}\geq 3. \tag{1}$$

Notice that because of the condition $x+y+z=0$, exactly one of the numbers xy, yz, zx is positive. Therefore, we may assume that $yz > 0$. By Cauchy-Schwarz inequality we have:

$$\frac{(2y+1)^2}{2y^2+1}+\frac{(2z+1)^2}{2z^2+1}\geq\frac{2(y+z+1)^2}{y^2+z^2+1}.$$

Using the condition $x+y+z=0$ and then $yz>0$ we obtain:

$$\frac{2(y+z+1)^2}{y^2+z^2+1}=\frac{2(x-1)^2}{x^2-2yz+1}\geq\frac{2(x-1)^2}{x^2+1}.$$

Going back to our inequality we have:

$$\frac{(2x+1)^2}{2x^2+1}+\frac{2(x-1)^2}{x^2+1}-3=\frac{2x^2(x-1)^2}{(2x^2+1)(x^2+1)}\geq 0.$$

This proves the required inequality. It is clear that, assuming that $yz>0$, the equality holds if and only if $x=1$ and $y=z=-1/2$.

Second solution. We again discard the case $xyz=0$. Observe that

$$2x^2+1=\frac{4}{3}x^2+\frac{2}{3}x^2+1=\frac{4}{3}x^2+\frac{2}{3}(y+z)^2+1\leq\frac{4}{3}x^2+\frac{4}{3}\left(y^2+z^2\right)+1,$$

by Cauchy-Schwarz. Equality holds if and only if $y=z$. Then we have

$$\frac{(2x+1)^2}{2x^2+1}\geq\frac{3\left(2x+1\right)^2}{4\left(x^2+y^2+z^2\right)+3},$$

with equality if either $y=z$, or $x=-\frac{1}{2}$. Adding up all similar inequalities yields

$$\sum\frac{(2x+1)^2}{2x^2+1}\geq 3\sum\frac{\left(2x+1\right)^2}{4\left(x^2+y^2+z^2\right)+3}=3,$$

which proves (1). Since $x = y = z$ implies that all three numbers are zero (case discarded), we deduce that the equality holds if two of the numbers equal $-\frac{1}{2}$ and the third equals 1.

28.3. Notice first that if $x < y < z$ are elements of a good set, then $y = xp^a$ and $z = yq^b$, where a, b are positive integers. Then $z = xp^a q^b$. But z/x is a power of a prime, hence $p = q$. This shows that the elements of a good set are among the members of a geometric sequence whose ratio is a prime number. Clearly, a largest good set T_0 has the form $a < ap < ap^2 < \cdots < ap^n$, where p is a prime number and n is the greatest exponent which appears in the prime representations of all elements of S. Of course, this maximum can be realized for several prime numbers p. The maximum length of such a set is $n+1$.

Let $S = T_1 \cup T_2 \cup \cdots \cup T_m$ be a partition of S with bad sets. Then, $T_i \cap T_0$ is a singleton, for all $i = 1, 2, \ldots, n$. Since every element of T_0 should appear in some T_i, $1 \le i \le n$, it follows that $m \ge n+1$. We will show that there exists a partition for which the lower bound $n+1$ can be realized. Indeed, if one takes

$$T_i = \{x \in S | \sum_{p\,\text{prime}} e_p(x) \equiv i (\operatorname{mod} n+1)\},$$

for all $i = 0, 1, \ldots, n+1$, then we obtain a partition with $n+1$ subsets. Here, $e_p(x)$ represents the exponent of the prime p in the prime decomposition of the integer number x.

28.4. We denote the vertices of the hexagon by $A_0, A_1, A_2, A_3, A_4, A_5$. From now on all indices should be taken modulo 6. For all $i = 0, 1, \ldots, 5$ we denote B_i the intersection of the lines A_iA_{i+1} and $A_{i+2}A_{i+3}$. In this way, one obtains the triangles $B_0B_2B_4$ and $B_1B_3B_5$ which overlap to define the hexagon $A_0A_1A_2A_3A_4A_5$. To show that the area of one of the triangles $B_0B_2B_4$ or $B_1B_3B_5$ is $\ge 3/2$ it is sufficient to prove that the total area of the triangles $A_{i+1}B_iA_{i+2}$ is at least 1. Indeed, if it is so, one of the sums

$$[A_1B_0A_2] + [A_3B_2A_4] + [A_5B_4A_0]$$

$$[A_2B_1A_3] + [A_4B_3A_5] + [A_0B_5A_1]$$

is not less than 1/2 and adding the area of the hexagon we obtain the result.

Denote by B_i' the reflection of the point B_i in the midpoint of the side $A_{i+1}A_{i+2}$ to obtain the parallelograms $B_iA_{i+1}B_i'A_{i+2}$, for all $i = 0, 1, \ldots, 5$. It is enough to prove that the six triangles $A_{i+1}B_iA_{i+2}$ cover the hexagon.

To do this, we reflect the points A_1, A_3, A_5 in the midpoints of the segments A_0A_2, A_2A_4, A_4A_0 respectively, to obtain the corresponding points A_1', A_3', A_5'. The hexagon is cut into the parallelograms $A_{2i}A_{2i+1}A_{2i+2}A_{2i+1}'$, $i = 0, 1, 2$ and the triangle $A_1'A_3'A_5'$, which can be degenerate.

It is clear that each parallelogram $A_{2i}A_{2i+1}A_{2i+2}A'_{2i+1}$ is covered by the pair of triangles $A_{2i}B'_{2i+5}A_{2i+1}$, $A_{2i+1}B'_{2i}A_{2i+2}$. The proof is complete if we prove that at least one of these pairs of triangles cover the triangle $A'_1A'_3A'_5$. For this, it is sufficient to prove that $A_{2i}B'_{2i+5} \geq A_{2i}A'_{2i+5}$ and $A_{2j+2}B'_{2j} \geq A_{2j+2}A'_{2j+3}$ for some indices $i, j \in \{0, 1, 2\}$. For the first inequality, we notice that

$$A_{2i}B'_{2i+5} = A_{2i+1}B_{2i+5}, A_{2i}A'_{2i+5} = A_{2i+4}A_{2i+5},$$

for all $i \in \{0, 1, 2\}$,

$$\frac{A_1B_5}{A_4A_5} = \frac{A_0B_5}{A_5B_3} \text{ and } \frac{A_3B_1}{A_0A_1} = \frac{A_2A_3}{A_0B_5}.$$

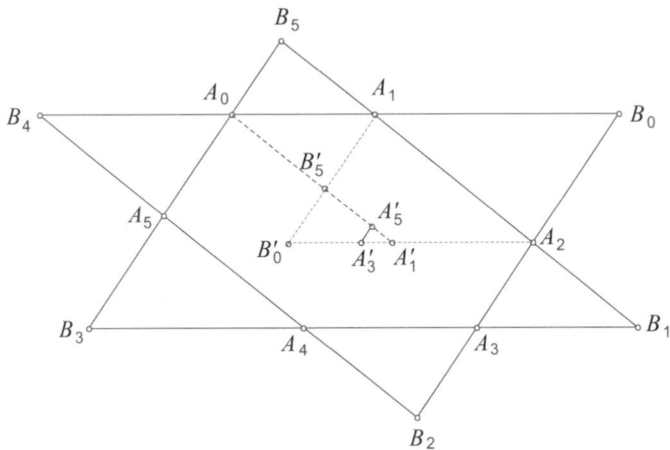

Figure 28.2

So, we obtain

$$\prod_{i=0}^{2} \frac{A_{2i}B'_{2i+5}}{A_{2i}A'_{2i+5}} = 1.$$

In the same way one obtains:

$$\prod_{j=0}^{2} \frac{A_{2j+2}B'_{2j}}{A_{2j+2}A'_{2j+3}} = 1.$$

The conclusion follows from these equalities.

Second solution. As in the previous solution, we show that the sum S of areas of all triangles $A_{i+1}B_iA_{i+2}$ is no less than 1.

First, observe that the six triangles $A_{i+1}B_iA_{i+2}$ and the two triangles $B_0B_2B_4$, $B_1B_3B_5$ are all similar. Therefore, we have

$$\frac{[A_0A_1B_5]}{[B_1B_3B_5]} = \left(\frac{A_1B_5}{B_1B_5}\right)^2, \frac{[A_1A_2B_0]}{[B_1B_3B_5]} = \left(\frac{A_1A_2}{B_1B_5}\right)^2, \frac{[A_2A_3B_1]}{[B_1B_3B_5]} = \left(\frac{A_2B_1}{B_1B_5}\right)^2.$$

Adding up, we obtain

$$\begin{aligned}\frac{[A_0A_1B_5] + [A_1A_2B_0] + [A_2A_3B_1]}{[B_1B_3B_5]} &= \left(\frac{A_1B_5}{B_1B_5}\right)^2 + \left(\frac{A_1A_2}{B_1B_5}\right)^2 + \left(\frac{A_2B_1}{B_1B_5}\right)^2 \\ &\geq \frac{1}{3}\left(\frac{A_1B_5}{B_1B_5} + \frac{A_1A_2}{B_1B_5} + \frac{A_2B_1}{B_1B_5}\right)^2 = \frac{1}{3}.\end{aligned}$$

It follows that

$$[A_0A_1B_5] + [A_1A_2B_0] + [A_2A_3B_1] \geq \frac{1}{3}[B_1B_3B_5].$$

Similarly, we have

$$[A_3A_4B_2] + [A_4A_5B_3] + [A_5A_0B_4] \geq \frac{1}{3}[B_0B_2B_4],$$

and we deduce that

$$S \geq \frac{1}{3}([B_1B_3B_5] + [B_0B_2B_4]) = \frac{1}{3}(S+2).$$

Obviously, this implies $S \geq 1$, as desired.

The 29th BMO

The twenty-ninth Balkan Mathematical Olympiad for high-school students was held between April 26th and May 2nd , 2012, in Antalya, Turkey. The number of participating countries was 21: 11 member countries (Albania, Bosnia and Herzegovina, Bulgaria, Cyprus, Former Yugoslav Republic of Macedonia, Greece, Republic of Moldova, Republic of Montenegro, Romania, Serbia and Turkey) and 10 invited countries (Afghanistan, Azerbaijan,France, Indonesia, Italy, Kazakhstan, Saudi Arabia, Tajikistan, Turkmenistan and United Kingdom.) Additionally, a second Turkish team took part as Turkey B.

Problems

29.1. Let A, B and C be points lying on a circle Γ with centre O. Assume that $\angle ABC > 90°$. Let D be the point of intersection of the line AB with the line perpendicular to AC at C. Let ℓ be the line through D which is perpendicular to AO. Let E be the point of intersection of ℓ with the line AC and F be the point of intersection of Γ with ℓ that lies between D and E. Prove that the circumcircles of triangles BFE and CFD are tangent at F.

(Romania)

29.2. Prove that

$$\sum_{cyc}(x+y)\sqrt{(z+x)(z+y)} \geq 4(xy+yz+zx),$$

for all positive real numbers x, y, z.

(Saudi Arabia)

29.3. Let n be a positive integer. Let $P_n = \{2^n, 2^{n-1}3, 2^{n-2}3^2, \ldots, 3^n\}$. For each subset X of P_n we write S_X for the sum of all elements of X, with the understanding that $S_\emptyset = 0$, where $\emptyset$ is the empty set. Suppose that y is a real number such that $0 \leq y \leq 3^{n+1} - 2^{n+1}$. Prove that there is a subset Y of P_n such that $0 \leq y - S_Y < 2^n$.

(United Kingdom)

29.4. Let $\mathbb{N}$ be the set of positive integers. Find all functions $f : \mathbb{N} \longrightarrow \mathbb{N}$ such that the following conditions hold:

(i) $f(n!) = f(n)!$, for all positive integers n,

(ii) $m - n$ divides $f(m) - f(n)$, whenever m and n are distinct positive integers.

(Saudi Arabia)

Solutions

29.1. We will use the following easy to prove property: given the triangle ABC, the line BD is tangent to its circumcircle at B if and only if $\angle BAD = \angle DBC$ (see Figure 29.1).

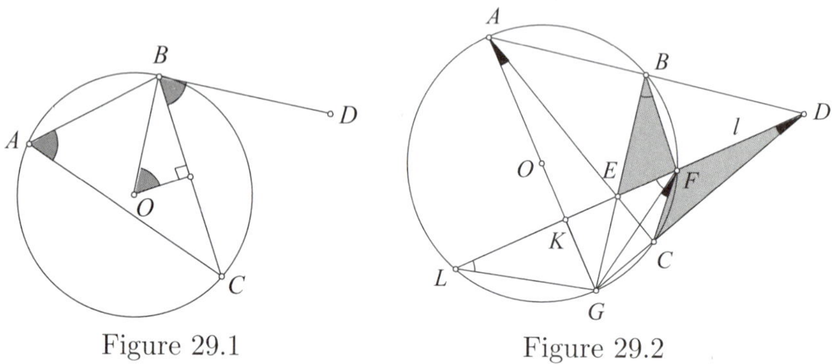

Figure 29.1 Figure 29.2

Let $\ell \cap AO = K$ and L the second intersection point of ℓ and Γ. Let G be the point of Γ diametrically opposed to A. Since $\angle ACD = 90^\circ$ it follows that D, C, G are collinear and AC is altitude in the triangle ADG. It follows that E is orthocenter in the triangle ADG and therefore G, E, B are collinear.

Since $\angle GFL = \angle GLF = \angle GBF$ it follows that GF is tangent to the circumcircle of the triangle BEF at F. Since $\angle GFC = \angle GAC = \angle GDF$ it follows that GF is tangent to the circumcircle of the triangle CDF at F. Since the two circles have a common tangent at F, they are tangent at F.

Second solution. We use the same picture as in the first solution. Let us consider the inversion of center D and ratio the power ρ of D with respect to Γ. Then $\rho = DB \cdot DA = DF \cdot DL = DC \cdot DG$. The triangles DBE and DKA are similar, so we have the proportionality: $DB/DE = DK/DA$. It follows that $\rho = DE \cdot DK$. This shows that the inversion maps the points B, F, C, E into the points A, L, G, K, respectively. Therefore the image by inversion of the circle through the points B, F, E is the circle circumscribed to the triangle ALK and the transformation by inversion of the circle through the points D, C, F is the line GL. The line GL is perpendicular to AL and AL is the diameter of the circle circumscribed to the triangle AKL. It follows that GL is tangent to that circle. This shows that the circles circumscribed to the triangles BFE and CFD have only one common point.

29.2. Since x, y, z are positive real numbers we can denote $y+z = a^2, z+x = b^2$ and $x + y = c^2$, where a, b, c are again positive real numbers. Then we can express:

$$x = 1/2(-a^2 + b^2 + c^2),\ y = 1/2(a^2 - b^2 + c^2) \text{ and } z = 1/2(a^2 + b^2 - c^2).$$

Plugging these formulas in the given inequality, we obtain the equivalent inequality:

$$\sum_{cyc} a^2bc \geq 2\sum_{cyc} a^2b^2 - \sum_{cyc} a^4.$$

First solution. Starting from the obvious inequality $(a+b)^2 > c^2$ we multiply it by $(a-b)^2$ to obtain $(a^2-b^2)^2 \geq (ca-bc)^2$. After computing the square and arranging the terms we obtain the inequality:

$$2abc^2 \geq 2a^2b^2 + b^2c^2 + c^2a^2 - a^4 - b^4.$$

Similarly one obtains the inequalities:

$$2a^2bc \geq 2b^2c^2 + c^2a^2 + a^2b^2 - b^4 - c^4,$$

$$2ab^2c \geq 2c^2a^2 + a^2b^2 + b^2c^2 - c^4 - a^4.$$

By adding these three inequalities one obtains the desired inequality.

Second solution. The inequalities $a + b > c$, $b + c > a$ and $c + a > b$ show that a, b, c are the sides of a triangle. The required inequality involves a symmetric polynomial in the variables a, b, c. Using the fundamental symmetric polynomials $S_1 = a + b + c$, $S_2 = ab + bc + ca$ and $S_3 = abc$, it can be written as

$$(a+b+c)^4 - 4(a+b+c)^2(ab+bc+ca) + 9(a+b+c) \geq 0.$$

After canceling the factor $a + b + c$ we obtain

$$(a+b+c)^3 - 4(a+b+c)(ab+bc+ca) + 9abc \geq 0.$$

It is known that in every triangle we have the identities: $a + b + c = 2s$, $ab + bc + ca = s^2 + r^2 + 4Rr$ and $abc = 4Rrs$, where s is the semi-perimeter,

R is the circumradius and r is the inradius. Then, the required inequality becomes:

$$8s^3 - 8s(s^2 + r^2 + 4Rr) + 36Rrs \geq 0.$$

After canceling similar monomials and simplifying rs, it reduces to Euler inequality: $R \geq 2r$.

Third solution. From Cauchy-Schwarz inequality we have:

$$\sqrt{z+x}\sqrt{z+y} \geq z + \sqrt{xy}$$

and similar inequalities obtained by cyclic permutations. By adding these inequalities we obtain:

$$\sum_{cyc}(x+y)\sqrt{(z+x)(z+y)} \geq \sum(x+y)(z+\sqrt{xy}) = 2\sum xy + \sum(x+y)\sqrt{xy}.$$

By the AM-GM inequality, $\sum(x+y)\sqrt{xy} \geq 2\sum xy$. This ends the proof.

29.3. Let $\varphi : \mathbb{R} \longrightarrow \mathbb{R}$ be the contraction of ratio $1/2^n$, $\varphi(x) = x/2^n$. The set P_n is mapped into a the set

$$P'_n = \left\{1, \frac{3}{2}, \left(\frac{3}{2}\right)^2, \ldots, \left(\frac{3}{2}\right)^n\right\}$$

and similarly, all numbers y and S_Y are divided by 2^n. For the simplicity we denote $a = 3/2$. Then the problem can be translated into the following: for every real number y, $0 \leq y \leq 1 + a + a^2 + \cdots + a^n$, there exists a subset $Y \subset P'_n$, such that $0 \leq y - S_Y < 1$. We will prove this by induction on n. For $n = 1$ it is obvious.

Suppose that the statement is true for n. Let y be a number $0 \leq y \leq 1 + a + a^2 + \cdots + a^{n+1}$. If $y \leq 1 + a + a^2 + \cdots + a^n$, using the induction hyphotesis, we choose the set Y as a subset of $P'_n \subset P'_{n+1}$. It remains the case $y > 1 + a + \cdots + a^n$. We claim that $y > a^{n+1}$. Indeed, we have

$$y > \frac{a^{n+1} - 1}{a - 1} = \frac{\left(\frac{3}{2}\right)^{n+1} - 1}{\frac{3}{2} - 1} = 2\left(\frac{3}{2}\right)^{n+1} - 2 > \left(\frac{3}{2}\right)^{n+1} = a^{n+1},$$

since $\left(\frac{3}{2}\right)^{n+1} > 2$, for all $n \geq 1$.

Then we have $0 < y - a^{n+1} \leq 1 + a + a^2 + \cdots + a^n$, and we use again the induction hyphotesis. In this way, we find a set $X \subset P'_n$ such that $0 \leq (y - a^{n+1}) - S_X < 1$. Taking $Y = \{a^{n+1}\} \cup X$ we have $0 \leq y - S_Y < 1$.

29.4. Assume that the constant function $f(n) = k$, for all n, is a solution. The condition (ii) is obviously verified and from condition (i) we have $k = k!$.

This is possible only for $k = 1$ or $k = 2$. So, we have two constant functions which are solutions of the problem. It is also obvious that the identity function verifies both conditions, giving another solution. We will prove that only these are possible solutions.

For every solution f we have $f(1) \in \{1, 2\}$ and $f(2) \in \{1, 2\}$. Let $k > 2$ be a fixed point of f. Then $f(k!) = f(k)! = k!$, showing that $k!$ is also a fixed point. This shows that $k, k!, (k!)!, ...$ is an increasing sequence of fixed points of f. So, we may assume that in this case, there exists an increasing sequence $a_1, a_2, \ldots, a_n, \ldots$ of fixed points.

First Solution. Assume that f has a fixed point $k \geq 3$. Let n be an arbitrary positive integer. Since $n - k$ divides $f(n) - f(k) = f(n) - k$ it follows $n - k | f(n) - n$. Now, taking the sequence $(a_i)_i$ we obtain that $f(n) = n$, for all n.

Assume that f has no fixed points ≥ 3. Then $f(3) \neq 3$. From $4 = 3! - 2$ and $3! - 2 | f(3)! - f(2)$ we have $f(3) \in \{1, 2\}$. To summarize, $f(1), f(2), f(3) \in \{1, 2\}$. Let $n > 3$ be an arbitrary integer. From $3 | n! - 3$ and $n! - 3 | f(n)! - f(3)$ it follows that $3 \not| f(n)!$. So, $f(n) \in \{1, 2\}$, for all $n \in \mathbb{N}$.

Since we assumed that f is not constant, there are distinct numbers m, n such that $f(m) = 1$ and $f(n) = 2$. If m, n are not consecutive we have $|n - m| > 1$ and $m - n | f(n) - f(m) = 1$; this is a contradiction. If they are always consecutive we take, for example, m and $m + 3$, to obtain the same contradiction.

Second Solution. Assume that f has no fixed points k, $k > 2$. Let $p \geq 5$ be a prime number. From Wilson's Theorem we have $p | (p-1)! + 1 = (p-1)(p-2)! + 1$, and so $p | (p-2)! - 1$. From condition (ii) one obtains $p | f(p-2)! - f(1)$. Since $f(1) \in \{1, 2\}$ and $p - 2 \geq 3$ it follows that $f(p-2) < p$. Since p does not divide $(p-1)! - 1$ and does not divide $(p-1)! - 2$ it follows that $f(p-2) \leq p-2$. Also, since f has no fixed points > 3, $f(p-2) < p-2$. Now, from $p-3 = (p-2) - 1$ we have $p - 3 | f(p-2) - f(1) < p - 3$. This is a contradiction, except when $f(p-2) = f(1) \in \{1, 2\}$. To end the proof, let n be an arbitrary integer. Then $(p-2) - n | f(n) - f(1)$, for all primes $p \geq 5$. This shows that $f(n) = f(1)$.

The 30th BMO

Although originally scheduled to take place in Albania, the thirtieth Balkan Mathematical Olympiad for high-school students was held in Agros, Cyprus, from June 28th to July 3th. The number of the participating countries was 14: 10 member countries (Albania, Bosnia and Herzegovina, Bulgaria, Cyprus, Greece, F.Y.R. of Macedonia, Moldova, Montenegro, Romania and Serbia) and 6 invited countries (Kazakhstan, United Kingdom, Azerbaijan, Tajikistan, Turkmenistan and Italy).

Problems

30.1. In a triangle ABC, the excircle ω_a opposite A touches AB at P and AC at Q, and the excircle ω_b opposite B touches BA at M and BC at N. Let K be the projection of C onto MN, and let L be the projection of C onto PQ. Show that the quadrilateral $MKLP$ is cyclic.

(Bulgaria)

30.2. Determine all positive integers x, y and z such that

$$x^5 + 4^y = 2013^z.$$

(Serbia)

30.3. Let S be the set of positive real numbers. Find all functions $f\colon S^3 \to S$ such that, for all positive real numbers x, y, z and k, the following three conditions are satisfied:

(a) $xf(x,y,z) = zf(z,y,x)$,
(b) $f(x,ky,k^2z) = kf(x,y,z)$,
(c) $f(1,k,k+1) = k+1$.

(United Kingdom)

30.4. In a mathematical competition, some competitors are friends; friendship is mutual, that is to say when A is a friend of B, then also B is a friend of A. We say that $n \geq 3$ different competitors $A_1, A_2, \ldots, A_n$ form a *weakly-friendly* *cycle* if A_i is not a friend of A_{i+1}, for $1 \leq i \leq n$ ($A_{n+1} = A_1$), and there are no other pairs of non-friends among the components of this cycle.

The following property is satisfied:

for every competitor C and every weakly-friendly cycle $\mathcal{S}$ of competitors not including C, the set of competitors D in $\mathcal{S}$ which are not friends of C has at most one element.

Prove that all competitors of this mathematical competition can be arranged into three rooms, such that every two competitors in the same room are friends.

(Serbia)

Solutions

30.1. Let S and T be the projections of points M and P on PQ and MN, respectively (see Figure 30.1). Denote by H the intersection between MS and PT and observe that triangles HTM and HSP are similar, hence

$$\frac{HS}{HT} = \frac{HP}{HM}.$$

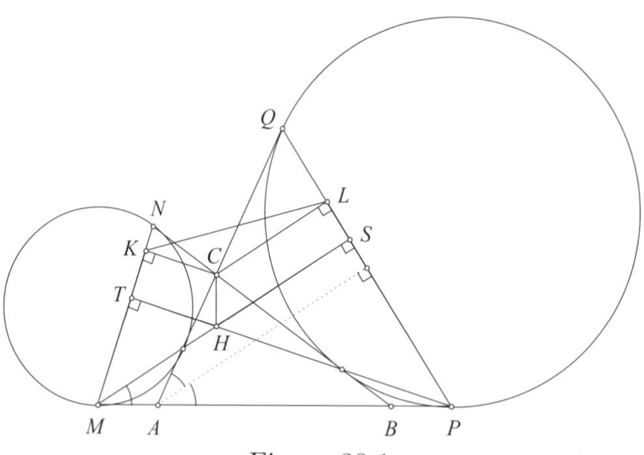

Figure 30.1

Since the line MS is parallel to the interior angle bisector of $\angle A$, we have $\angle HMP = \frac{\angle A}{2}$. Similarly, $\angle HPM = \frac{\angle B}{2}$. Using the sine law in triangle HMP we obtain

$$\frac{HP}{HM} = \frac{\sin\frac{A}{2}}{\sin\frac{B}{2}}.$$

In triangle CQL we have $\angle CQL = 90^\circ - \frac{A}{2}$, and $CQ = s - b$ (see Appendix, The incircle and the excircles of a triangle). Therefore

$$CL = CQ \sin\left(90^\circ - \frac{A}{2}\right) = (s-b)\cos\frac{A}{2}.$$

Analogously,

$$CK = (s-a)\cos\frac{B}{2}.$$

It follows that

$$\frac{CL}{CK} = \frac{(s-b)\cos\frac{A}{2}}{(s-a)\cos\frac{B}{2}}$$

Since we have

$$r = (s-a)\tan\frac{A}{2} = (s-b)\tan\frac{B}{2}$$

(Appendix, The incircle and the excircles of a triangle) we deduce that

$$\frac{CL}{CK} = \frac{HS}{HT}.$$

But HS is parallel to CL, and HT is parallel to CK. We deduce that triangles HST and CLK are homothetic, therefore KL is parallel to ST. Obviously, $MTSP$ is a cyclic quadrilateral, hence $MKLP$ is cyclic, as well.

Second solution. Suppose the lines MN and PQ intersect at D. We claim that CD is perpendicular to AB (see Figure 30.2).

If this were true, then

$$\angle CDK = 90^\circ - \angle SMB = \angle SBM = \frac{B}{2}.$$

But $CLDK$ is a cyclic quadrilateral, hence $\angle CDK = \angle CLK$. Finally, in the quadrilateral $MPLK$, we have

$$\angle KMP + \angle KLP = 90^\circ - \frac{B}{2} + 90^\circ + \frac{B}{2} = 180^\circ,$$

and it follows that $MPLK$ is cyclic, as desired.

Let us now prove our claim. Let I_a, I_b be the centers of the excircles, and suppose the line I_aI_b intersects MN and PQ at E and F, respectively (see Figure 30.3).

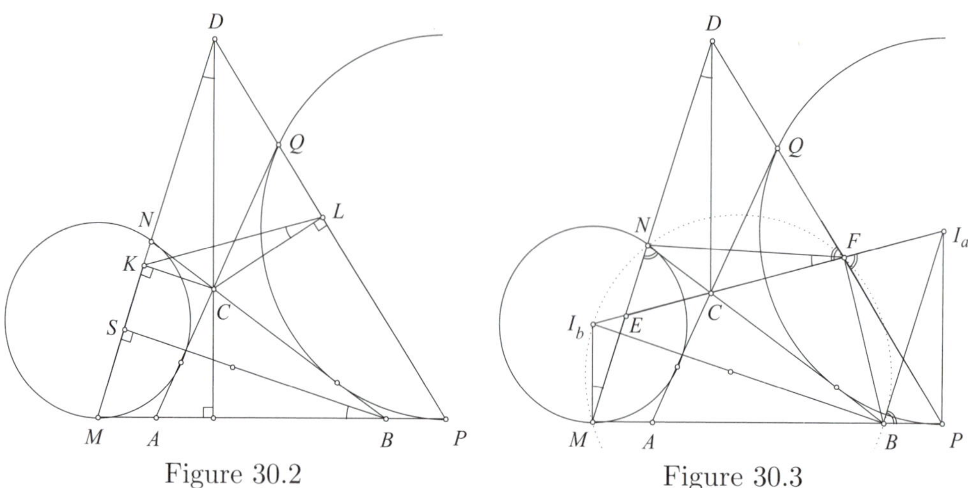

Figure 30.2 Figure 30.3

We have

$$\angle FPB = \angle FI_aB = 90^\circ - \frac{A}{2},$$

hence the quadrilateral BPI_aF is inscribed in a circle $\mathcal{C}$. We deduce that

$$\angle BFI_b = BPI_a = 90^\circ,$$

therefore the line segment BI_b is a diameter of $\mathcal{C}$, and it is not difficult to see that points M and N lie on $\mathcal{C}$, as well.

Finally, some angle chasing: we have

$$\angle MNB = 90^\circ - \frac{B}{2},$$

and

$$\angle DFE = \angle I_aFP = \angle I_aBP = 90^\circ - \frac{B}{2},$$

so

$$\angle MNB = \angle DFE,$$

which implies that $ANCF$ is cyclic. But then

$$\angle CDN = \angle CFN = \angle I_bFN = \angle I_bMK,$$

hence CD is parallel to I_bM.

Clearly I_bM is perpendicular to AB, so CD is also perpendicular to AB, as claimed.

Observation. The second solution shows that the problem is, in fact, a disguise of an old geometry gem. Let ABC be a triangle and consider its excircles. The lines defined by the tangency points of each excircle to the extensions of triangle's sides determine the triangle $A'B'C'$ (see Figure 30.4). Then the lines AA', BB', and CC' are the altitudes of the triangle ABC.

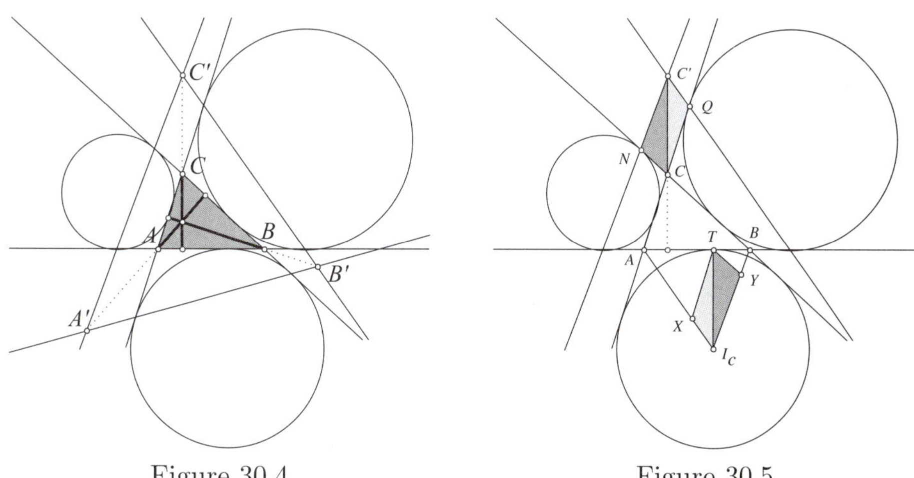

Figure 30.4 Figure 30.5

A nice proof of this result[9] is obtained by considering points X on AI_c and Y on BY_c (here I_c obviously denotes the center of the excircle tangent to AB -see Figure 30.5-) such that TX is parallel to AC and TY is parallel to BC. Then we have $\angle BTY = \angle B$ and $\angle TBY = 90^\circ - \angle B/2$, hence

$$\angle TYB = 180^\circ - (\angle BTY + \angle TBY) = 90^\circ - \angle B/2$$

as well, so the triangle TBY is isosceles, with $TB = TY = s-a$ (see Appendix, The incircle and the excircles of a triangle).

Similarly, $TA = TX = s-b$. We claim that the quadrilaterals TYI_cX and $CNC'Q$ are congruent. Indeed, we have $TY = CN = s-a$, $TX = CQ = s-b$, $\angle YTX = \angle BCA = \angle QCN$, $\angle TXI_c = \angle CQC' = 90^\circ + \angle A/2$, and $\angle TYI_c = \angle CNC'$. Therefore, $\angle C'CN = \angle I_cTY = 90^\circ - \angle B$, and it follows that $C'C$ is perpendicular to AB.

30.2. Taking the equation modulo 11 yields

$$x^5 + 4^y \equiv 0 \pmod{11}.$$

If 11 does not divide x, then, by Fermat's little theorem, we have $x^{10} \equiv 1 \pmod{11}$, hence

$$\left(x^5 - 1\right)\left(x^5 + 1\right) \equiv 0 \pmod{11}.$$

It follows that $x^5 \equiv 0, 1, -1 \pmod{11}$. Checking the powers of 4, we see that $4^y \equiv 1, 3, 4, 5, 9 \pmod{11}$, and we conclude that if the equation has a solution,

[9] The authors wish to thank Professor Dan Fulea for communicating his solution.

then we must have $x^5 \equiv -1$ and $4^y \equiv 1 \pmod{11}$. The latter happens if and only if y is divisible by 5, so let $y = 5t$, for some positive integer t.

Denoting $4^t = u$, the equation rewrites as

$$(x+u)\left(x^4 - x^3 \cdot u + x^2 \cdot u^2 - x \cdot u^3 + u^4\right) = 2013^z.$$

We claim that

$$\gcd\left(x+u, x^4 - x^3 \cdot u + x^2 \cdot u^2 - x \cdot u^3 + u^4\right) = 1.$$

Indeed, let p be a prime dividing $x+u$. Observe that $p \in \{3, 11, 61\}$, since $2013 = 3 \cdot 11 \cdot 61$. Then $u \equiv -x \pmod{p}$, hence

$$x^4 - x^3 \cdot u + x^2 \cdot u^2 - x \cdot u^3 + u^4 \equiv 5x^4 \pmod{p}.$$

We cannot have $5x^4 \equiv 0 \pmod{p}$, otherwise it follows that $p|x$, hence $p|u = 4^t$, that is, $p = 2$, a contradiction.

Since $x^5 + 4^y \equiv x^5 + 1 \equiv 0 \pmod{3}$, it follows that $x + 1 \equiv 0 \pmod{3}$, that is, 3 divides $x + 4^t = x + u$. We have two cases.

If $x + u = 3^z$, then

$$2013^z = x^5 + u^5 < (x+u)^5 = 3^{5z} = 243^z,$$

a contradiction.

If $x + u = 33^z$, then

$$2013^z = x^5 + u^5 = 2 \cdot \frac{x^5 + u^5}{2} > 2 \cdot \left(\frac{x+u}{2}\right)^5 = \frac{33^{5z}}{16} = \frac{39135393^z}{16},$$

again a contradiction. We conclude that the equation has no solution in positive integers.

30.3. We claim that

$$f(tx, ty, tz) = f(x, y, z),$$

fo all x, y, z, t in S. Indeed, let $k = \sqrt{t}$; we then have

$$\begin{aligned} f(tx, ty, tz) &= f\left(k^2x, k^2y, k^2z\right) = kf\left(k^2x, ky, z\right) \\ &= \frac{1}{kx}k^2xf\left(k^2x, ky, z\right) = \frac{1}{kx}zf\left(z, ky, k^2x\right) \\ &= \frac{z}{x}f(z, y, z) = f(x, y, z). \end{aligned}$$

Therefore, if $x, y, z > 0$, then

$$f(x, y, z) = f\left(1, \frac{y}{x}, \frac{z}{x}\right).$$

Let $k, m > 0$ such that $km = \frac{y}{x}$ and $k^2(m+1) = \frac{z}{x}$. Then

$$f\left(1, \frac{y}{x}, \frac{z}{x}\right) = f\left(1, km, k^2(m+1)\right) = kf(1, m, m+1) = k(m+1) = \frac{y}{x} + k.$$

It remains to observe that k is a positive root of the equation

$$k\frac{y}{x} + k^2 = \frac{z}{x},$$

which rewrites as

$$xk^2 + yk - z = 0,$$

and we obtain

$$k = \frac{-y + \sqrt{y^2 + 4xz}}{2x},$$

and then

$$f(x, y, z) = \frac{y}{x} + \frac{-y + \sqrt{y^2 + 4xz}}{2x} = \frac{y + \sqrt{y^2 + 4xz}}{2x},$$

for all $x, y, z > 0$. It is not difficult to check that f indeed satisfies the three given conditions.

Second solution. Observe that

$$\begin{aligned} f(x, y, z) &= f\left(x, \sqrt{z} \cdot \frac{y}{\sqrt{z}}, \sqrt{z}^2 \cdot 1\right) = \sqrt{z} f\left(x, \frac{y}{\sqrt{z}}, 1\right) \\ &= \frac{\sqrt{z}}{x} \cdot x f\left(x, \frac{y}{\sqrt{z}}, 1\right) = \frac{\sqrt{z}}{x} f\left(1, \frac{y}{\sqrt{z}}, x\right) \\ &= \frac{1}{x} f(1, y, xz). \end{aligned}$$

We look for $k, m > 0$ such that $km = y$ and $k^2(m+1) = xz$. We find that k is the positive root of the equation

$$k^2 + yk - xz = 0,$$

hence

$$k = \frac{-y + \sqrt{y^2 + 4xz}}{2}.$$

Then

$$\begin{aligned} \frac{1}{x} f(1, y, xz) &= \frac{1}{x} f\left(1, km, k^2(m+1)\right) = \frac{k}{x} f(1, m, m+1) \\ &= \frac{k}{x}(m+1) = \frac{km}{x} + \frac{k}{x} = \frac{y}{x} + \frac{-y + \sqrt{y^2 + 4xz}}{2x} \\ &= \frac{y + \sqrt{y^2 + 4xz}}{2x}. \end{aligned}$$

30.4. It is natural to consider the graph G in which the competitors are represented by vertices and non-friendships by edges. Thus, a *weakly-friendly cycle* is a cycle of G formed by the vertices $A_1, A_2, \dots, A_n$, in which A_i is adjacent to A_{i+1}, for $1 \le i \le n$ ($A_{n+1} = A_1$) and no other pair of vertices among $A_1, A_2, \dots, A_n$ are connected by an edge. Call this a *cordless cycle*. Then the condition of the problem restates as follows: for every vertex C and every cordless cycle $\mathcal{S}$, not including C, at most one vertex of $\mathcal{S}$ is adjacent to C. We have to prove that one can color the vertices of G with three colors such that two vertices having the same color are not connected by an edge.

We will prove that if the conditions of the problems are fulfilled, then the graph contains at least one vertex with degree 2 or less. If so, then the assertion goes by induction on the number of vertices: skipping the trivial base case, choose a vertex V of minimal degree (which is 2 or less) and observe that the graph $G - \{V\}$ still satisfies the given condition. Color $G - \{V\}$ with three colors as demanded. Finally, since V is connected with at most 2 other vertices, color V with the remaining color (or one of the remaining colors).

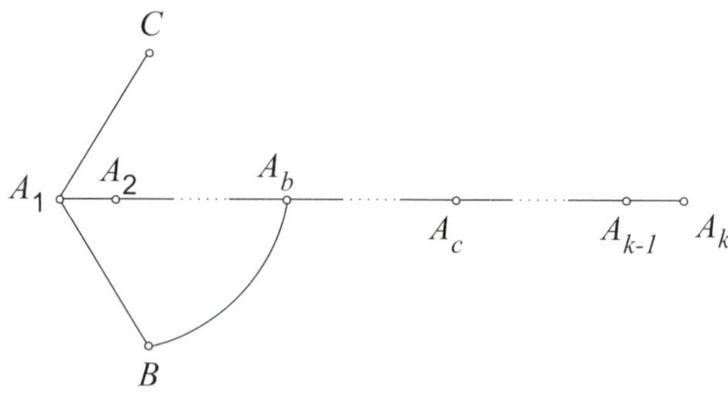

Figure 30.6

Now, the proof of our claim: suppose, by way of contradiction, that all vertices have the degree at least 3. Consider $\mathcal{P}$, the longest cordless path $A_1, A_2, \dots, A_k$ (A_i is adjacent to A_{i+1}, for $1 \le i \le k-1$ and no other pair of vertices among $A_1, A_2, \dots, A_k$ are connected by an edge).

Since $\deg(A_1) \ge 3$, there exist vertices $B, C \notin \mathcal{P}$, connected to A_1 (see Figure 30.6). Each of these two vertices must be adjacent to some vertex in $\mathcal{P}$, otherwise the maximality of $\mathcal{P}$ is contradicted. Suppose that A_b is the vertex in $\mathcal{P} - \{A_1\}$ adjacent to B and closest to A_1, and define similarly A_c. Furthermore, suppose that $b \le c$. But then $A_1, A_2, \dots, A_c, C$ is a cordless cycle in G not containing B and B is adjacent to two different vertices of this cycle: A_1 and A_b. This contradiction ends our proof.

Part II

Supplementary Problems

Geometry

Problems

G.1. A line passing through the center O of an equilateral triangle ABC intersects the circumcircles of the triangles OAB, OBC, and OCA at K, L, and M. Prove that

$$OK^2 + OL^2 + OM^2 = 2AB^2.$$

G.2. Let M be a point inside the equilateral triangle ABC and let A', B', and C' be its projections on the sides BC, CA, and AB, respectively. Denote by $r_1, r_2, r_3, r'_1, r'_2$, and r'_3 the inradii of the triangles MAC', MBA', MCB', MAB', MBC', and MCA'. Prove that

$$r_1 + r_2 + r_3 = r'_1 + r'_2 + r'_3.$$

G.3. Let $ABCD$ be a convex quadrilateral and let M, N be points on the sides AB and BC. The line segments DM and DN intersect AC at K and L and the lines BK and BL intersect the sides AD and CD at R and S, respectively. Suppose that $AK = KL = LC$ and

$$[ADM] = [CDN] = \frac{1}{4}[ABCD].$$

Prove that

$$[ABR] = [BCS] = \frac{1}{4}[ABCD].$$

G.4. On each side of the triangle ABC a regular $n-$gon is constructed in the exterior on the triangle and sharing the side with the triangle. Find the values of n for which the centers of the three $n-$gons are the vertices of an equilateral triangle.

G.5. Let O be a point in the interior of the triangle ABC. The lines OA, OB, and OC intersect the sides BC, CA, and AB at the points D, E, and F, respectively. The same lines intersect EF, FD, and DE at the points K, L, and M. Prove that

$$\frac{KA}{KD} \cdot \frac{LB}{LE} \cdot \frac{MC}{MF} \geq 1.$$

G.6. Find the maximum number of points that can be chosen in the interior of a regular hexagon with side length 1 such that all mutual distances between the points are at least $\sqrt{2}$.

G.7. The side lengths of the obtuse triangle ABC are three consecutive odd integers and $11\cos\angle B = 13\cos\angle C$. Prove that $OI = \frac{2}{3}\sqrt{ab}$ (O is the circumcenter and I is the incenter of the triangle).

G.8. Let $A_1A_2\ldots A_n$ be a regular polygon. Find all points P in the polygon's plane with the property: the squares of distances from P to the polygon's vertices are consecutive terms of an arithmetic sequence.

G.9. Let ABC be an arbitrary triangle. If M is a point in the triangle's plane, let M_1, M_2, M_3 be the reflections of M across the triangle's sides and let M^* be the centroid of the triangle $M_1M_2M_3$. Find M such that $M = M^*$.

Solutions

G.1. Suppose that the vertices of the triangle ABC are oriented counterclockwise (see Figure S.1). Observe that rotating clockwise the circumcircle of BOC around O with 120° we obtain the circumcircle of AOB. This rotation sends the line segment OL to the line segment OL', where L' is a point on the circumcircle of AOB such that $\angle KOL' = 60^\circ$. Similarly, if M' is a point on the same circle such that $\angle M'OK = 60^\circ$, then $OM' = OM$.

Because the side of an equilateral triangle inscribed in a circle with radius R equals $R\sqrt{3}$, the problem can be restated as follows: if four distinct points O, K, L, and M are chosen on a circle with radius R such that $\angle KOL = \angle MOK = 60^\circ$, then

$$OK^2 + OL^2 + OM^2 = 6R^2.$$

It is not difficult to see that the length of a chord which subtends an arc of measure α in a circle with radius R equals $2R\sin\frac{\alpha}{2}$. Therefore, denoting, for instance, the measure of the arc OL with α, we have to prove that

$$4R^2\left(\sin^2\frac{\alpha}{2} + \sin^2\frac{\alpha+120^\circ}{2} + \sin^2\frac{\alpha+240^\circ}{2}\right) = 6R^2.$$

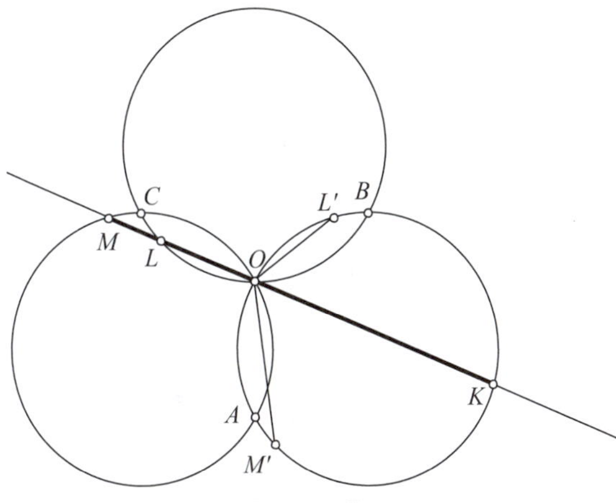

Figure S.1

But $2\sin^2 x = 1 - \cos 2x$, so the above equality becomes

$$\cos\alpha + \cos(\alpha + 120^\circ) + \cos(\alpha + 240^\circ) = 0,$$

which is easy to prove. Indeed, we have

$$\cos(\alpha + 120^\circ) = -\frac{1}{2}\cos\alpha - \frac{\sqrt{3}}{2}\sin\alpha$$
$$\cos(\alpha + 240^\circ) = -\frac{1}{2}\cos\alpha + \frac{\sqrt{3}}{2}\sin\alpha$$

hence

$$\cos(\alpha + 120^\circ) + \cos(\alpha + 240^\circ) = -\cos\alpha,$$

as needed.

Second solution. We use complex numbers. With no loss of generality, we can assume that the affixes of the points K, L, and M are the cubic roots of the unity, that is, $1, \varepsilon_1$, and ε_2. If z is the affix of the point O, then $|z| = 1$, since O must lie on the unit circle (see Figure S.2).

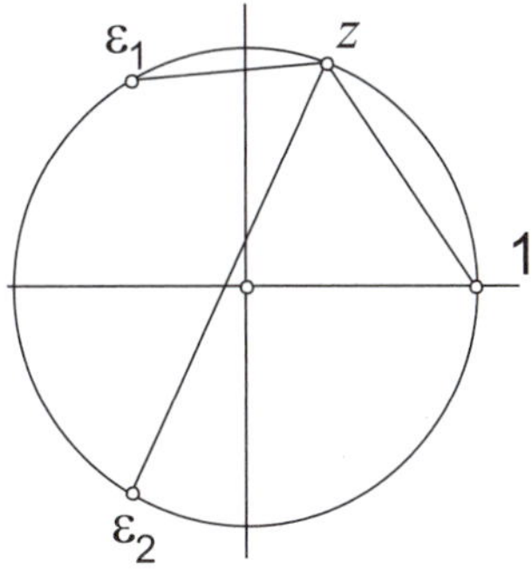

Figure S.2

We have to prove that

$$|z-1|^2+|z-\varepsilon_1|^2+|z-\varepsilon_2|^2=6.$$

Because ε_1 and ε_2 are the roots of the equation $x^2+x+1=0$, we have $\varepsilon_1+\varepsilon_2=-1$ and $\varepsilon_1\varepsilon_2=1$. Moreover, $\overline{\varepsilon}_1=\varepsilon_2$ and $\overline{\varepsilon}_2=\varepsilon_1$. Using this, we have

$$\begin{aligned}
6&=|z-1|^2+|z-\varepsilon_1|^2+|z-\varepsilon_2|^2\\
&=(z-1)(\overline{z}-1)+(z-\varepsilon_1)(\overline{z}-\overline{\varepsilon}_1)+(z-\varepsilon_2)(\overline{z}-\overline{\varepsilon}_2)\\
&=(z-1)(\overline{z}-1)+(z-\varepsilon_1)(\overline{z}-\varepsilon_2)+(z-\varepsilon_2)(\overline{z}-\varepsilon_1)\\
&=3|z|^2-z(1+\varepsilon_1+\varepsilon_2)-\overline{z}(1+\varepsilon_1+\varepsilon_2)+3\\
&=3|z|^2+3,
\end{aligned}$$

which is obvious.

Observation. Using again complex numbers, a more general statement can be proved: if $A_0A_1\ldots A_{n-1}$ is a regular polygon inscribed in the unit circle and P is an arbitrary point on the unit circle, then

$$PA_0^2+PA_1^2+\ldots+PA_{n-1}^2=2n.$$

Indeed, let ε_k, $0\le k\le n-1$, be the affixes of the points A_k and z the affix of P. We can assume that ε_k are the roots of the polynomial X^n-1. Then

$$\begin{aligned}
PA_0^2+PA_1^2+\ldots+PA_{n-1}^2&=|z-\varepsilon_0|^2+|z-\varepsilon_1|^2+\ldots+|z-\varepsilon_{n-1}|^2\\
&=\sum(z-\varepsilon_k)(\overline{z}-\overline{\varepsilon}_k)\\
&=\sum\left(|z|^2-z\overline{\varepsilon}_k-\overline{z}\varepsilon_k+|\varepsilon_k|^2\right)\\
&=2n-\overline{z}\sum\varepsilon_k-z\overline{\left(\sum\varepsilon_k\right)}\\
&=2n,
\end{aligned}$$

since $|z|=|\varepsilon_k|=1$ and Viète's relations give $\sum\varepsilon_k=0$.

G.2. We use an additional result.

Lemma. Let XYZ be a right triangle, with $\angle X=90^\circ$ and side lengths x,y,z, respectively. If r is the inradius of XYZ, then $r=\frac{1}{2}(y+z-x)$.

Proof. It is known that $S=rp$, where S denotes the area of the triangle and p its semiperimeter. Therefore, we have

$$r=\frac{yz}{x+y+z}.$$

But $x^2 = y^2 + z^2$, so that

$$\frac{yz}{x+y+z} = \frac{yz\,(y+z-x)}{(y+z)^2 - x^2} = \frac{yz\,(y+z-x)}{2yz} = \frac{1}{2}\,(y+z-x)\,.$$

Applying the lemma in our problem, a short computation shows that the requested equality is equivalent to

$$AB' + BC' + CA' = AC' + BA' + CB'.$$

To prove this equality observe that

$$A'M^2 = BM^2 - A'B^2 = CM^2 - A'C^2,$$

and hence

$$BM^2 - CM^2 = A'B^2 - A'C^2 = (A'B - A'C) \cdot a,$$

where a is the side length of the triangle. We get the desired result by writing the other two similar equalities and adding them up.

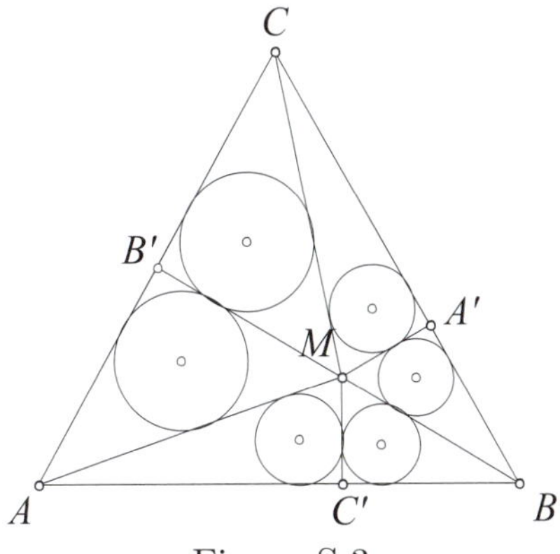

Figure S.3

Observation. The last argument is a particular case of the theorem of Carnot: let A', B, and C' be points on the sides BC, CA, and AB of an arbitrary triangle ABC; then the perpendiculars on the sides erected at A', B', and C' are concurrent if and only if

$$A'B^2 + B'C^2 + C'A^2 = A'C^2 + B'A^2 + C'B^2.$$

G.3. Suppose, for the beginning, that $AK = LC$ and $[ADM] = [CDN]$. We will prove that $[ABR] = [BCS]$. Observe that $[ADK] = [CDL]$, because $AK = LC$. Therefore, $[AKM] = [CLN]$, hence the points M and N are equally distanced from the line AC. It follows that MN is parallel to AC (see Figure S4). Let O be the intersection between the lines AC and BD. Applying Ceva's theorem in the triangles ABD and BCD yields

$$\frac{RD}{RA} \cdot \frac{OB}{OD} \cdot \frac{MA}{MB} = 1,$$
$$\frac{SD}{SC} \cdot \frac{OB}{OD} \cdot \frac{NC}{NB} = 1.$$

But $\frac{MA}{MB} = \frac{NC}{NB}$, therefore $\frac{RD}{RA} = \frac{SD}{SC}$, that is, RS is parallel to AC. Then $[ARK] = [CSL]$ and since $[ABK] = [BLC]$, we obtain $[ABR] = [BCS]$, as desired.

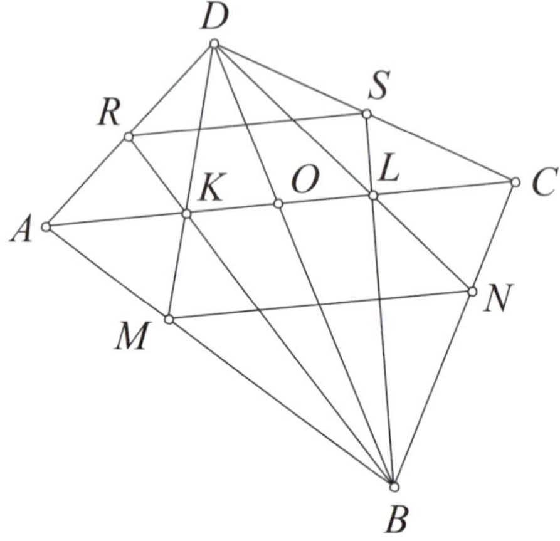

Figure S.4

Next, we prove that if $[ADM] = [CDN] = \frac{1}{4}[ABCD]$ and $AK = KL = LC$, then $ABCD$ is a parallelogram and M, N, S, and R are the midpoint of its sides. Indeed let

$$k = \frac{MB}{MA} = \frac{NB}{NC}.$$

Then

$$[MBD] = k\,[MAD]$$
$$[NBD] = k\,[NCD],$$

and hence

$$\begin{aligned}[ABCD] &= [ABD] + [BCD] \\ &= (1+k)\,([MAD] + [NCD]) \\ &= (1+k) \cdot \frac{1}{2}\,[ABCD].\end{aligned}$$

We obtain $k = 1$, so that M and N are the midpoints of AB and BC. Because $AK = KL$, MK is a midline in the triangle ABL, hence MK is parallel to BL. Similarly, LN is parallel to BK therefore $BKDL$ is a parallelogram. It follows that O is the midpoint of the line segment BD so we conclude that K is the centroid of the triangle ABD. This implies that R is the midpoint of the side AD. Analogously, S is the midpoint of CD. Finally, since O is the midpoint of KL, it is the midpoint of AC as well. Thus, the line segments AC and BD have the same midpoint, implying that $ABCD$ is a parallelogram (see Figure S.5).

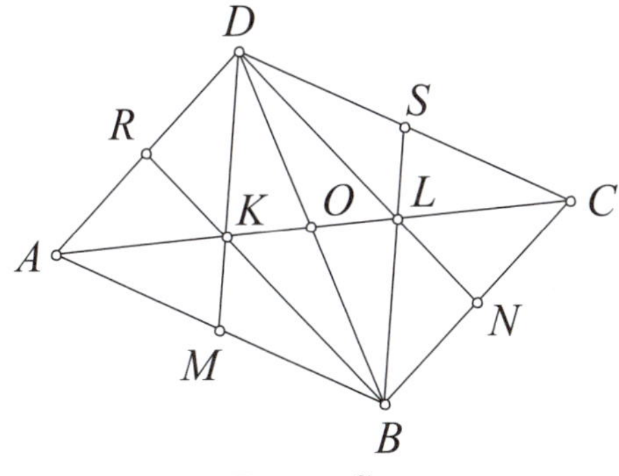

Figure S.5

It is not difficult to see that in this case

$$[ABR] = [BCS] = \frac{1}{4}[ABCD]$$

as desired.

G.4. We start with a known result.
Lemma. Let a, b, and c be the affixes of the noncollinear points A, B, C. The triangle ABC is equilateral if and only if

$$a + b\varepsilon + c\varepsilon^2 = 0,$$

where ε is either one of the nonreal cubic roots of the unity.
Proof. Suppose that the vertices A, B, C are arranged counterclockwise (as in Figure S.6). Then the triangle ABC is equilateral if and only if

$$a - c = (c - b)\,\varepsilon,$$

where $= \cos\frac{2\pi}{3} + i\sin\frac{2\pi}{3}$.

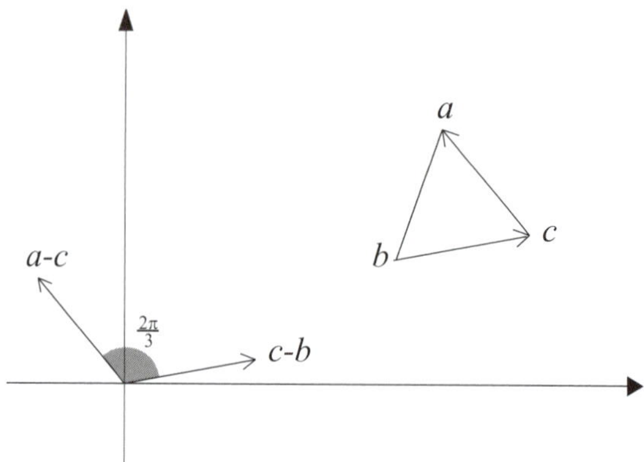

Figure S.6

The equality can be written as

$$a + b\varepsilon + c\,(-\varepsilon - 1) = 0.$$

But $\varepsilon^2 + \varepsilon + 1 = 0$, therefore the equality above becomes

$$a + b\varepsilon + c\varepsilon^2 = 0.$$

If A, B, C are arranged clockwise, then A, C, B are arranged counterclockwise, so that we obtain

$$a + c\varepsilon + b\varepsilon^2 = 0.$$

If $\varepsilon' = \cos\frac{4\pi}{3} + i\sin\frac{4\pi}{3}$ is the other nonreal cubic root of the unity, then $\varepsilon' = \varepsilon^2$ and $\varepsilon'^2 = \varepsilon^4 = \varepsilon$, therefore we have

$$a + b\varepsilon' + c\varepsilon'^2 = 0.$$

Returning to our problem, let

$$\omega = \cos\frac{2\pi}{n} + i\sin\frac{2\pi}{n},$$

and let a, b, c be the affixes of the triangle's vertices. If z is the affix of the center of the n-gon constructed on the side AB, then we must have

$$b - z = \omega\,(a - z)\,,$$

therefore

$$z = \frac{\omega a - b}{\omega - 1}.$$

We deduce that the three centers are the vertices of an equilateral triangle if and only if

$$\frac{\omega a - b}{\omega - 1} + \varepsilon\frac{\omega b - c}{\omega - 1} + \varepsilon^2\frac{\omega c - a}{\omega - 1} = 0,$$

where ε is either one of the nonreal cubic roots of the unity. Using $\varepsilon^3 = 1$, the latter becomes

$$\left(a + b\varepsilon + c\varepsilon^2\right)\left(\omega - \varepsilon^2\right) = 0.$$

It follows that either ABC is an equilateral triangle and $n \geq 3$ is an arbitrary integer or $n = 3$.

Observation. The case $n = 3$ is known as Napoleon's problem (see Figure S.7).

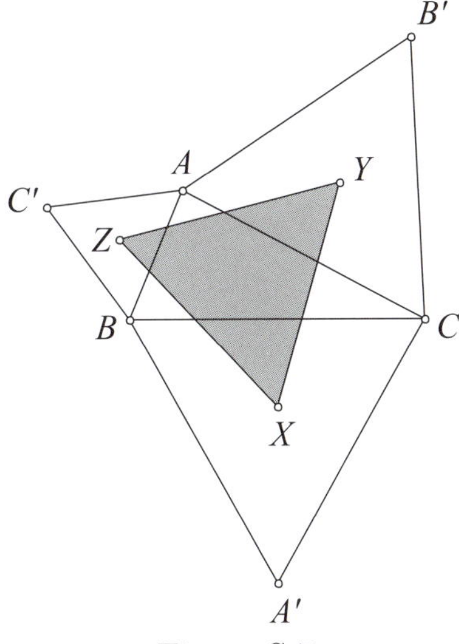

Figure S.7

G.5. We will use the following result.

Lemma. Let X, Y, and Z be points on the sides AB, AC, and BC of the triangle ABC such that the segments CX, BY, and AZ are concurrent. If AZ and XY intersect at T then

$$\frac{AT}{TZ} = \frac{1}{2}\left(\frac{AX}{XB} + \frac{AY}{YC}\right).$$

Proof. If

$$\frac{AX}{XB} = \frac{AY}{YC},$$

then XY is parallel to BC and hence

$$\frac{AT}{TZ} = \frac{AX}{XB} = \frac{AY}{YC} = \frac{1}{2}\left(\frac{AX}{XB} + \frac{AY}{YC}\right).$$

Suppose that XY intersects BC at U (see Figure S.8). Applying Menelaus' theorem in triangle ABZ yields

$$\frac{AT}{TZ} \cdot \frac{XB}{AX} \cdot \frac{UZ}{UB} = 1,$$

hence

$$\frac{AT}{TZ} = \frac{AX}{XB} \cdot \frac{UB}{UZ}.$$

Similarly,

$$\frac{AT}{TZ} = \frac{AY}{YC} \cdot \frac{UC}{UZ}.$$

Thus, we have to prove that

$$\frac{AX}{XB} \cdot \frac{UB}{UZ} + \frac{AY}{YC} \cdot \frac{UC}{UZ} = \frac{AX}{XB} + \frac{AY}{YC},$$

which is equivalent to

$$\frac{AY}{YC} \cdot \frac{CZ}{UZ} = \frac{AX}{XB} \cdot \frac{BZ}{UZ},$$

or

$$\frac{AY}{YC} \cdot \frac{CZ}{BZ} \cdot \frac{XB}{AY} = 1.$$

The latter follows from the theorem of Ceva applied in triangle ABC.

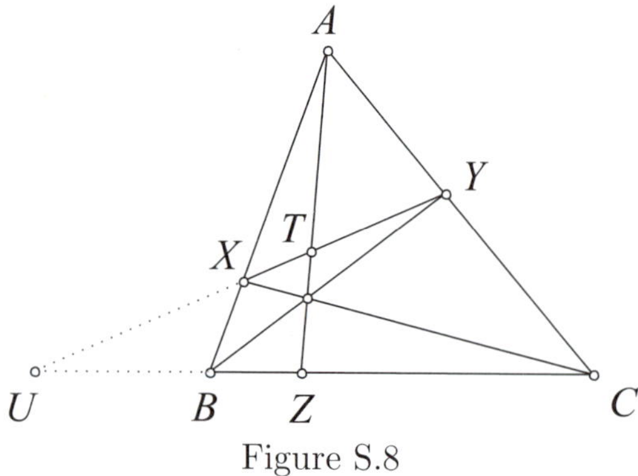

Figure S.8

Returning to our problem, denote

$$\frac{AF}{FB} = x, \ \frac{CE}{EA} = y, \ \frac{BD}{DC} = z.$$

Using the lemma and the AM-GM inequality, we obtain

$$\frac{KA}{KD} \cdot \frac{LB}{LE} \cdot \frac{MC}{MF} = \frac{1}{2}\left(x + \frac{1}{y}\right) \cdot \frac{1}{2}\left(y + \frac{1}{z}\right) \cdot \frac{1}{2}\left(z + \frac{1}{x}\right)$$
$$\geq \sqrt{\frac{x}{y}} \cdot \sqrt{\frac{y}{z}} \cdot \sqrt{\frac{z}{x}} = 1,$$

as needed. The equality holds if and only if $x = y = z = 1$ and this happens when O is the centroid of the triangle ABC.

G.6. Let O be the hexagon's center and let A, B be two of the points such that the angle$\angle AOB$ is minimal. The cosine law in the triangle OAB gives

$$\cos \angle AOB = \frac{OA^2 + OB^2 - AB^2}{2OA \cdot OB}.$$

Since A and B lie in the interior of the hexagon, we have $OA, OB < 1$. Moreover, $AB \geq \sqrt{2}$, so it results that $\cos \angle AOB < 0$, and hence $\angle AOB < 90°$. We deduce that there are no more than three points satisfying the given condition. To conclude, we give an example with three points lying in the interior of the hexagon such that all mutual distances are at least $\sqrt{2}$.

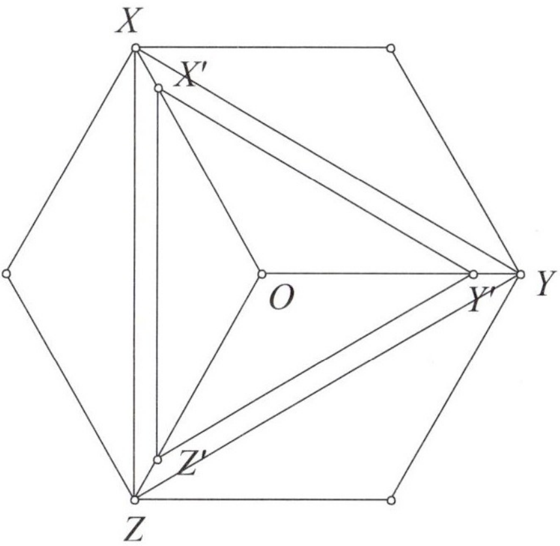

Figure S.9

Let X, Y, and Z be three nonadjacent vertices of the hexagon. A homothety centered at O with ratio $r = \sqrt{2/3}$ sends the points X, Y, Z to the points X', Y', Z', respectively (see Figure S.9). A short computation shows that the side length of the equilateral triangle $X'Y'Z'$ equals $\sqrt{2}$.

G.7. Since $\cos \angle B$ and $\cos \angle C$ have the same sign, we deduce that $\angle A$ is obtuse and $\angle B, \angle C$ are acute angles. Moreover, $\cos \angle B < \cos \angle C$, hence $a > c > b$. The cosine law gives

$$\cos A = \frac{b^2 + c^2 - a^2}{2bc} < 0,$$

therefore $b^2 + c^2 < a^2$. Setting $a = 2k + 1$, $b = 2k - 1$, and $c = 2k - 3$, the previous inequality yields $2 \leq k \leq 4$, hence the triple (a, b, c) is $(5, 3, 1)$, $(7, 5, 3)$, or $(9, 7, 5)$. Because $b + c > a$, we are left with only two candidates: $(7, 5, 3)$ and $(9, 7, 5)$. A short computation using the cosine law shows that only $(7, 5, 3)$ satisfies the equality $11 \cos \angle B = 13 \cos \angle C$. Therefore, $a = 7$, $b = 3$ and $c = 5$.

Recall the Euler triangle formula (see Appendix), that is: in every triangle we have

$$OI^2 = R^2 - 2Rr,$$

where R and r are the circumradius and the inradius. The rest is trivial: we use the well known formulae

$$S = pr = \frac{abc}{4R},$$

where S is the area and p is the semiperimeter of the triangle. Heron's formula yields

$$S = \frac{15\sqrt{3}}{4}$$

and then

$$R = \frac{7\sqrt{3}}{3}, \quad r = \frac{\sqrt{3}}{2}.$$

Plugging in (2) we obtain

$$OI = \frac{2}{3}\sqrt{21} = \frac{2}{3}\sqrt{ab},$$

as needed.

G.8. Given the triangle ABC, let us find the points P such that

$$2PA^2 = PB^2 + PC^2.$$

If D is the midpoint of the side BC, the theorem of the median gives

$$PB^2 + PC^2 = 2PD^2 + \frac{BC^2}{2}$$

and hence we have

$$PA^2 - PD^2 = \frac{BC^2}{4}.$$

Let P' be the projection of P on the line AD (see Figure S.10). Then

$$\frac{BC^2}{4} = PA^2 - PD^2 = P'A^2 - P'D^2,$$

so we deduce that P' is a fixed point. Therefore P lies on a line which is perpendicular to the median AD. Moreover, if O is the circumcenter of ABC, we have

$$OA^2 - OD^2 = OB^2 - OD^2 = BD^2 = \frac{BC^2}{4},$$

hence PP' passes through O.

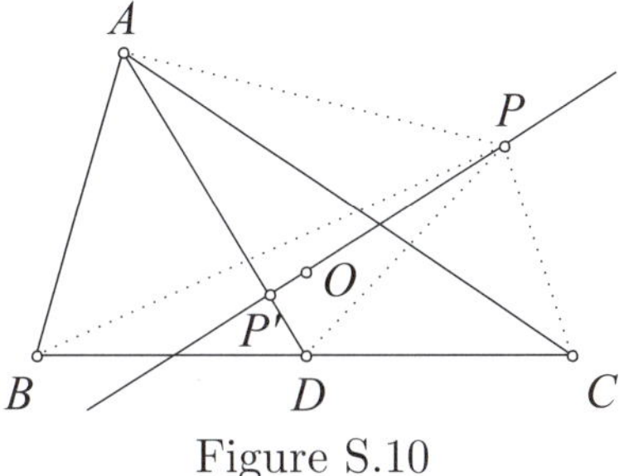

Figure S.10

We conclude that the locus of the point P is the perpendicular to the median AD which passes through the circumcenter O of the triangle ABC.

Returning to the problem, let us first consider the case $n = 3$. Applying the above result, we deduce that the locus of the points P such that PA_1^2, PA_2^2, and PA_3^2 are consecutive terms of an arithmetic sequence (in some order) is the reunion of three lines, each passing through the triangle's center and being parallel to one of the triangle's sides (see Figure S.11).

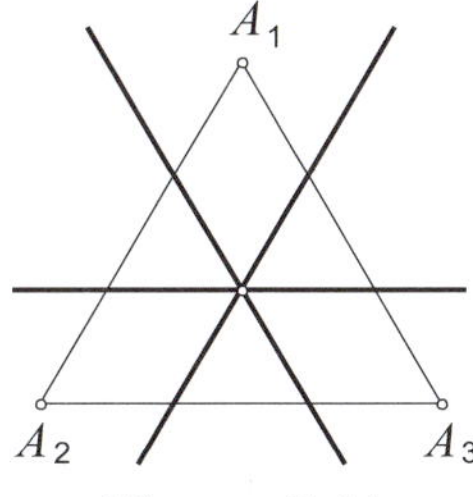

Figure S.11

In the case $n = 4$, consider the square $A_1A_2A_3A_4$ with center O, and draw lines through the opposite vertices and through the midpoints of the opposite sides, thus dividing the plane into eight regions (see Figure S.12). The region in which P lies determines the relative ordering of the distances PA_1, PA_2, PA_3, and PA_4. Suppose P lies in region R_1, as in Figure S.12. Then we have

$$PA_2 \leq PA_1 \leq PA_3 \leq PA_4,$$

and, unless $P = O$, the previous inequalities are strict ones.

From the hypothesis it results that

$$2PA_1^2 = PA_2^2 + PA_3^2$$

and

$$2PA_3^2 = PA_1^2 + PA_4^2.$$

Denoting by B and C the midpoints of the sides A_2A_3 and A_1A_4, we deduce that P belongs to the perpendicular dropped from O to A_1B, and to the perpendicular dropped from O to A_3C. Since A_1B is parallel to A_3C, the two perpendiculars coincide.

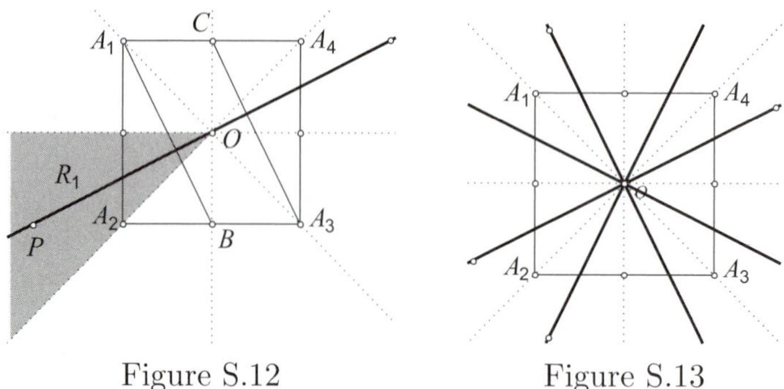

Figure S.12

Figure S.13

We conclude that the locus of point P is the reunion of four lines passing through O, as in Figure S.13.

If $n \geq 5$, we proceed in a similar way, but in this case the locus reduces to the point O, the center of the polygon. Examine Figure S.14 and observe that if P belongs to the marked region, then

$$PA_1 \leq PA_n \leq PA_2 \leq PA_{n-1}$$

and B_1A_n is not parallel to $B_{n-1}A_2$. Therefore the perpendiculars dropped from O to the lines B_1A_n and $B_{n-1}A_2$ intersect at O.

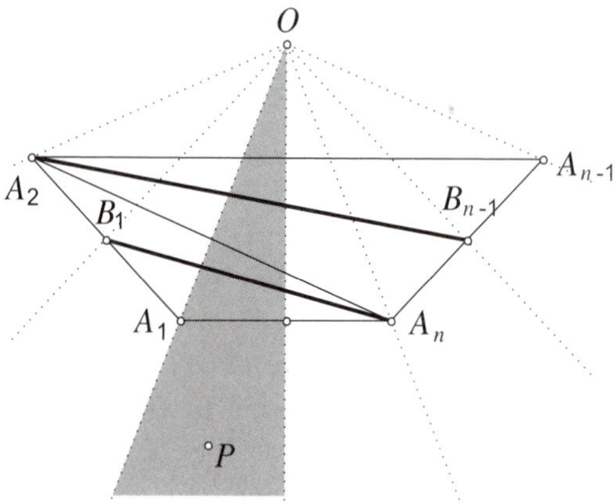

Figure S.14

G.9. A homothety centered at M with ratio $\frac{1}{2}$ sends the points M_1, M_2, and M_3 to the points A', B', and C', the projections of P to the triangle's sides (see Figure S.15).

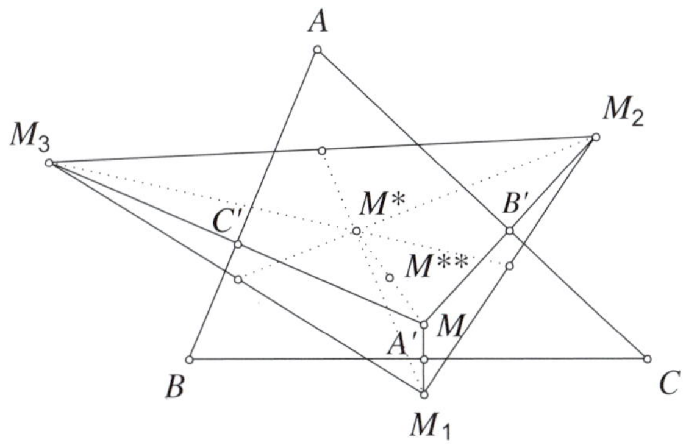

Figure S.15

The same homothety sends M^* to M^{**}, the centroid of triangle $A'B'C'$. It is clear that $M = M^*$ if and only if $M = M^{**}$.

Thus, we have to find the point M such that it is the centroid of the triangle determined by its projections on the sides of triangle ABC. Suppose MA' intersects $B'C'$ at N (see Figure S.16). Then

$$\frac{NC'}{NB'} = \frac{[NC'A']}{[NB'A']} = \frac{A'C' \cdot \sin\angle NA'C'}{A'B' \cdot \sin\angle NA'B'}. \tag{1}$$

The sine law applied in triangle $MA'C'$ gives

$$\frac{A'C'}{\sin\angle A'MC'} = \frac{MC'}{\sin\angle NA'C'}.$$

Since the quadrilateral $A'BC'M$ is cyclic (its opposite angles $\angle A'$ and $\angle C'$ are right angles) it follows that $\sin\angle A'MC' = \sin(180^\circ - \angle B) = \sin\angle B$. We deduce that

$$A'C' \cdot \sin\angle NA'C' = MC' \cdot \sin\angle B = MC' \cdot \frac{AC}{2R},$$

the latter following from the sine law in triangle ABC (R is the circumradius of ABC). Similarly, we obtain

$$A'B' \cdot \sin\angle NA'B' = MB' \cdot \frac{AB}{2R},$$

and hence (1) yields

$$\frac{NC'}{NB'} = \frac{MC'}{MB'} \cdot \frac{AC}{AB}.$$

If M is the centroid of $A'B'C'$ then $NC' = NB'$ and hence

$$\frac{MB'}{MC'} = \frac{AC}{AB}.$$

We deduce that M is the centroid of $A'B'C'$ if and only if the distances from M to the sides of ABC are proportional to the respective side lengths. We show that in this case M is the symmedian point of the triangle ABC i.e. the intersection point of the symmedians (a symmedian is the symmetric of a median with respect to the angle bisector).

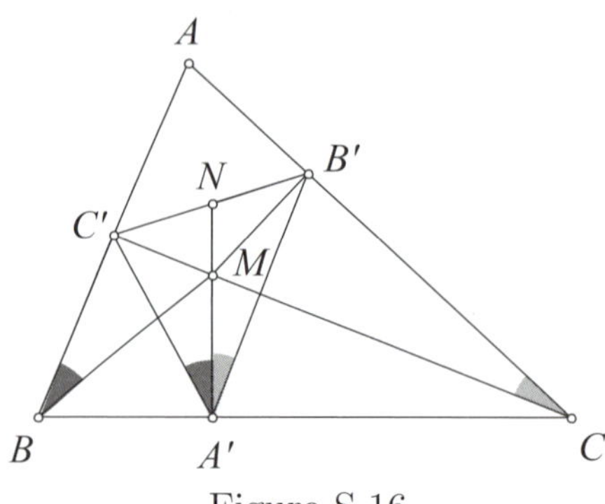

Figure S.16

Indeed, let M' be the reflection of M across the angle bisector AD, let B'', C'' be its projections on the sides AC, AB and let K the point where AM' intersects the side BC (see Figure S.17). Obvious symmetry reasons give

$$\frac{M'C''}{M'B''} = \frac{MB'}{MC'}$$

hence

$$\frac{M'C''}{M'B''} = \frac{AC}{AB},$$

and therefore

$$[ABM'] = [ACM'],$$

which can be written as

$$AM' \cdot BB_1 = AM' \cdot CC_1,$$

where B_1 and C_1 are the projections of the points B and C on the line AM'. We obtain $BB_1 = CC_1$ and then $BK = CK$, which shows that M' lies on the median AK. Using similar arguments we deduce that M' is the centroid of triangle ABC an hence M is the symmedian point.

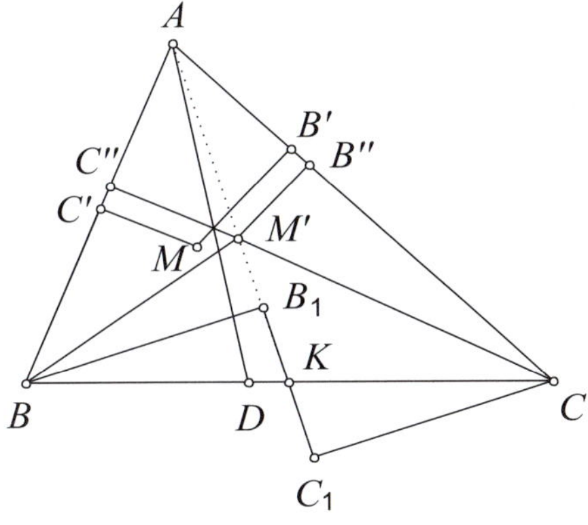

Figure S.17

Observation. The concurrence of the symmedians follows easily from the trigonometric form of Ceva's theorem. The symmedian point is also called Lemoine's point of the triangle.

Algebra and Number Theory

Problems

A.1. The sequence $(x_n)_{n\geq 1}$ is defined by $x_1 = 3$, $x_2 = 2$, and

$$x_{n-1}x_{n+1} = x_n^2 + 5,$$

for all $n \geq 2$. Prove that all terms of the sequence are positive integers.

A.2. Let x, y, z, and t be positive integers such that

$$\frac{x}{yz} + \frac{y}{zx} + \frac{z}{xy} = t.$$

Prove that t equals either 1 or 3.

A.3. Let $(x_n)_{n\geq 1}$ be a sequence defined by $x_1 = 1$ and

$$x_{n+1} = 1 + 2x_n,$$

for all n.

a) Find n for which x_n is divisible by 5.

b) Prove that there exists no positive integer $n \geq 2$ such that n divides x_n.

A.4. Let $\mathcal{F} = \{E_1, E_2, \ldots, E_n\}$ be a family of subsets of a given finite set with the following property: if $E \in \mathcal{F}$ and $E' \subset E$, then $E' \in \mathcal{F}$. Prove that there exists a permutation $\sigma : \mathcal{F} \to \mathcal{F}$ such that

$$E_i \cap \sigma(E_i) = \emptyset,$$

for each i, $1 \leq i \leq n$.

A.5. Let $n \geq 2$ be a positive integer and $p \geq 3$ be a prime number.

a) Prove that the polynomial

$$f(X) = X^n + 2^p$$

can be written as a product of two nonconstant integer polynomials if and only if n is divisible by p.

b) Prove that he polynomial

$$g(X) = X^n + 4$$

can be written as a product of two nonconstant integer polynomials if and only if n is divisible by 4.

A.6. Let $p < q$ be two prime numbers. Prove that the equation

$$\frac{1}{x} - \frac{1}{y} = \frac{1}{p} - \frac{1}{q}$$

has one, two or four solutions in positive integers.

Solutions

A.1. Observe that all terms are positive numbers. Subtracting the equations

$$x_{n-1}x_{n+1} = x_n^2 + 5$$
$$x_n x_{n+2} = x_{n+1}^2 + 5,$$

we obtain

$$x_n x_{n+2} - x_{n-1}x_{n+1} = x_{n+1}^2 - x_n^2,$$

which is equivalent to

$$\frac{x_{n+2} + x_n}{x_{n+1}} = \frac{x_{n+1} + x_{n-1}}{x_n}.$$

Iterating this, we deduce that

$$\frac{x_{n+2} + x_n}{x_{n+1}} = \frac{x_3 + x_1}{x_2} = 3,$$

and hence

$$x_{n+2} = 3x_{n+1} - x_n,$$

for all $n \geq 1$. Since x_1 and x_2 are integers, it follows inductively that all terms of the sequence are positive integers.

A.2. The given equation is equivalent to

$$x^2 + y^2 + z^2 = txyz,$$

and for given t, we consider this an equation with x, y, z as unknowns.

Suppose $t \geq 4$; we show that in this case the equation has no solutions in positive integers. Assume, by way of contradiction, that (a, b, c) is a solution such that $\max(a, b, c)$ is minimal. With no loss of generality, we may consider $a \geq b \geq c$. We have $a > b$, since otherwise

$$tabc = ta^2c \geq 4a^2 > a^2 + b^2 + c^2,$$

a contradiction. Also, we obtain

$$abc > b^2,$$
$$abc \geq c^2 + bc,$$

hence

$$2abc > b^2 + c^2 + bc.$$

We deduce that

$$a^2 + b^2 + c^2 = tabc \geq 4abc > 2b^2 + 2c^2 + 2bc,$$

and this yields

$$a^2 > (b + c)^2,$$

or, equivalently,

$$a > b + c.$$

Now, observe that if (a, b, c) is a solution, then $(tbc - a, b, c)$ is also a solution. Indeed, we have

$$(tbc - a)^2 + b^2 + c^2 - t(tbc - a)bc = a^2 + b^2 + c^2 - tabc = 0.$$

On the other hand,

$$tabc = a^2 + b^2 + c^2 < a^2 + ab + ac,$$

hence

$$tbc - a < b + c < a.$$

This shows that $\max(tbc - a, b, c) < \max(a, b, c)$ contradicting the choice of the solution (a, b, c).

Next, we show that if $t = 2$, the equation has no solution in positive integers. Suppose (a, b, c) is a solution such that $a + b + c$ is minimal. From

$$a^2 + b^2 + c^2 = 2abc,$$

we observe that either all three numbers are even, or one is even and the other two are odd. In the second case, we obtain a contradiction by noticing that the left hand side of the equality is congruent to 2 (mod 4) while the right hand side is not. It follows that a, b, and c are even and setting $a = 2a_1$, $b = 2b_1$, $c = 2c_1$ we obtain

$$a_1^2 + b_1^2 + c_1^2 = 2a_1b_1c_1,$$

that is, (a_1, b_1, c_1) is also a solution, with $a_1 + b_1 + c_1 < a + b + c$. This contradiction proves our claim.

To conclude, observe that for $t = 1$ the equation has the solution $(3,3,3)$ and for $t = 3$, the solution $(1,1,1)$.

A.3. a) A simple induction shows that $x_n = 2^n - 1$, for all $n \geq 1$. Since

$$2^2 \equiv -1 \pmod 5,$$

we deduce that

$$2^n - 1 \equiv 0 \pmod 5)$$

if and only if n is divisible by 4.

b) Let $a, N > 1$ be positive integers such that $\gcd(a, N) = 1$. The least positive integer k such that $a^k \equiv 1 \pmod N$ is called the order of a modulo N (we denote $k = \text{ord } a \pmod N$). It is well known that if m is a positive integer such that $a^m \equiv 1 \pmod N$, then k divides m.

Suppose that $n \geq 2$ divides $2^n - 1$. Then

$$2^n \equiv 1 \pmod n.$$

Let $p > 1$ be the least prime divisor of n. Then we have

$$2^n \equiv 1 \pmod p$$

and Fermat's little theorem yields

$$2^{p-1} \equiv 1 \pmod p.$$

Let k be the order of 2 modulo p. It follows that k divides both n and $p - 1$ and since p is the least prime divisor of n, it follows that $k = 1$, thus $p = 1$, a contradiction.

A.4. We use induction on the number $K_n$of elements of $E_1 \cup E_2 \cup \ldots \cup E_n$. Consider the nontrivial case $K_n = 2$,and let $\cup E_i = \{a, b\}$. Then $n \leq 4$; moreover, if $\{a, b\} \in \mathcal{F}$, we have

$$\mathcal{F} = \{E_1, E_2, E_3, E_4\},$$

where $E_1 = \emptyset$, $E_2 = \{a\}$, $E_3 = \{b\}$, and $E_4 = \{a, b\}$. Then the permutation $\sigma : \mathcal{F} \to \mathcal{F}$ defined by

$$\sigma(E_1) = E_4, \ \sigma(E_2) = E_3, \ \sigma(E_3) = E_2, \ \sigma(E_4) = E_1$$

clearly has the desired property. If $\{a,b\} \notin \mathcal{F}$ then

$$\mathcal{F} = \{E_1, E_2, E_3\},$$

where $E_1 = \emptyset$, $E_2 = \{a\}$, and $E_3 = \{b\}$. In this case, any permutation $\sigma : \mathcal{F} \to \mathcal{F}$ has the requested property.

Now, let $x \in \cup E_i$ and define

$$\mathcal{F}_1 = \{E - \{x\} \mid E \in \mathcal{F} \text{ and } x \in E\}$$
$$\mathcal{F}_2 = \{E \mid E \in \mathcal{F} \text{ and } x \notin E\}.$$

Observe that both $\mathcal{F}_1$ and $\mathcal{F}_2$ have the property: if $E \in \mathcal{F}_i$ and $E' \subset E$, then $E' \in \mathcal{F}_i$, hence from the induction hypothesis, we can find permutations $\sigma_i : \mathcal{F}_i \to \mathcal{F}_i$ such that

$$E \cap \sigma_i(E) = \emptyset,$$

for all $E \in \mathcal{F}_i$. We define $\sigma : \mathcal{F} \to \mathcal{F}$ as follows:

$$\sigma(E) = \begin{cases} \sigma_1(E - \{x\}), & \text{if } x \in E; \\ \sigma_2(E) \cup \{x\}, & \text{if } x \notin E \text{ and } \sigma_2(E) \cup \{x\} \in \mathcal{F}; \\ \sigma_2(E), & \text{in all other cases.} \end{cases}$$

It is not difficult to see that σ is well defined and that for all $E \in \mathcal{F}$ we have

$$E \cap \sigma_i(E) = \emptyset.$$

A.5. Let $x_1, x_2, \ldots, x_n$ be the roots of the polynomial $f(X)$. It is not difficult to see that they have equal absolute values and since Viète's relations yield

$$x_1 x_2 \ldots x_n = (-1)^n 2^p,$$

we obtain

$$|x_1| = |x_2| = \ldots = |x_n| = 2^{\frac{p}{n}}.$$

Now, suppose f can be written as a product of two nonconstant integer polynomials:

$$f(X) = f_1(X) f_2(X),$$

where

$$f_1(X) = a_m X^m + \ldots + a_0,$$

and

$$f_2(X) = b_s X^s + \ldots + b_0.$$

We obtain $a_m b_s = 1$, hence either $a_m = b_s = 1$, or $a_m = b_s = -1$. Let $x_1, \ldots, x_m$ be the roots of f_1 and $x_{m+1}, \ldots, x_n$ the roots of f_2. Then

$$x_1 x_2 \ldots x_m = (-1)^m \frac{a_0}{a_m},$$

so that the product $x_1x_2\ldots x_m$ is an integer number. Taking the absolute values, we obtain that

$$N = 2^{\frac{pm}{n}}$$

is an integer. This is possible only if

$$\frac{pm}{n} = t$$

is an integer. Since $m < n$, we must have $t < p$, therefore from the equality

$$pm = nt$$

we deduce that p divides n (recall that p is a prime number). Observe that we did not use the condition $p \geq 3$.

Conversely, if p is an odd prime which divides n, write $n = pk$ and observe that

$$\begin{aligned} X^{pk} + 2^p &= \left(X^k\right)^p + 2^p \\ &= \left(X^k + 2\right)\left(X^{k(p-1)} + X^{k(p-2)} \cdot 2 + \ldots + 2^{p-1}\right). \end{aligned}$$

b) The argument used before works as well: if g factors, then n must be an even integer. Moreover, if $n = 2k$, then g factors either as

$$g(X) = \left(X^k + a_{k-1}X^{k-1} + \ldots + 2\right)\left(X^k + b_{k-1}X^{k-1} + \ldots + 2\right), \quad (1)$$

or as

$$g(X) = \left(X^k + a_{k-1}X^{k-1} + \ldots - 2\right)\left(X^k + b_{k-1}X^{k-1} + \ldots - 2\right). \quad (2)$$

The key observation is that the factors of g are irreducible polynomials in $\mathbb{Z}[X]$. Take, for instance

$$h(X) = X^k + a_{k-1}X^{k-1} + \ldots + 2.$$

Its roots are roots of g as well, and hence their absolute value is $\sqrt[n]{4} = \sqrt[k]{2}$. Suppose h factors into integer nonconstant polynomials

$$h(X) = (X^s + \ldots + a)\left(X^t + \ldots + b\right),$$

with $s, t \geq 2$. Then $ab = 2$, so we can assume $|a| = 1$ and $|b| = 2$. If $x_1, x_2, \ldots, x_s$ are the roots of $X^s + \ldots + a$, then

$$x_1x_2\ldots x_s = (-1)^s a,$$

therefore

$$2^{\frac{s}{k}} = 1,$$

a contradiction. Finally, assume that k is odd, and suppose g factors as in (1). We have then

$$\begin{aligned} g(-X) &= \left(-X^k + a_{k-1}X^{k-1} + \ldots + 2\right)\left(-X^k + b_{k-1}X^{k-1} + \ldots + 2\right) \\ &= \left(X^k - a_{k-1}X^{k-1} - \ldots - 2\right)\left(X^k - b_{k-1}X^{k-1} - \ldots - 2\right). \end{aligned}$$

But $g(-X) = g(X)$, since g is an even function. It follows that g factors in two different ways as a product of irreducible polynomials, and this is a contradiction.

Conversely, if $n = 4k$, for some positive integer k, then

$$\begin{aligned} X^{4k} + 4 &= X^{4k} + 4X^{2k} + 4 - 4X^{2k} \\ &= \left(X^{2k} + 2\right)^2 - \left(2X^k\right)^2 \\ &= \left(X^{2k} + 2X^k + 2\right)\left(X^{2k} - 2X^k + 2\right), \end{aligned}$$

so g is reducible in $\mathbb{Z}[X]$.

A.6. The given equation is equivalent to

$$pq(y - x) = xy(q - p), \tag{1}$$

and since p and q are different prime numbers, they are both relatively prime to $q - p$. Therefore, each of p and q divides the product xy. We distinguish several cases.

- p divides only x, q divides only y.

  Write $x = px'$, $y = qy'$, with $\gcd(x', q) = \gcd(y', p) = 1$. Plugging in (1) yields

  $$qy' - px' = x'y'(q - p),$$

  and we deduce that x' and y' divide each other, hence $x' = y'$. This implies $x' = y' = 1$, leading to the solution $(x, y) = (p, q)$.

- p divides both x and y, q divides only y.

  Write $x = px'$, $y = pqy'$, with $\gcd(x', q) = 1$. We obtain

  $$qy' - x' = x'y'(q - p),$$

  and, again, $x' = y'$. We obtain $x' = \frac{q-1}{q-p}$, and hence the solution $(x, y) = \left(p\frac{q-1}{q-p}, pq\frac{q-1}{q-p}\right)$ if and only if $\frac{q-1}{q-p}$ is an integer number.

- p divides only y, q divides both x and y.

  Write $x = qx'$, $y = pqy'$, with $\gcd(x', p) = 1$. From (1) we get

$$py' - x' = x'y'(q-p),$$

  and, again, x' and y' divide each other, yielding $x' = y' = \frac{p-1}{q-p}$. We obtain the solution $(x, y) = \left(q\frac{p-1}{q-p}, pq\frac{p-1}{q-p}\right)$ if an only if $\frac{p-1}{q-p}$ is an integer number.

- p and q divide only y.

  In this case we write $y = pqy'$, with $\gcd(x, p) = \gcd(x, q) = 1$. It follows that

$$pqy' - x = xy'(q-p).$$

  Once again, x and y' divide each other, so $x = y' = \frac{pq-1}{q-p}$. We deduce that $(x, y) = \left(\frac{pq-1}{q-p}, pq\frac{pq-1}{q-p}\right)$ is a solution if and only if $\frac{pq-1}{q-p}$ is an integer number.

It is not difficult to check that in each of the remaining cases we reach a contradiction.

Finally, observe that

$$\frac{q-1}{q-p} - \frac{p-1}{q-p} = 1,$$

hence $\frac{q-1}{q-p}$ and $\frac{p-1}{q-p}$ are simultaneously integer numbers. Also,

$$\frac{pq-1}{q-p} = -q + (q+1)\frac{q-1}{q-p},$$

therefore if $\frac{q-1}{q-p}$ is an integer, then $\frac{pq-1}{q-p}$ is an integer as well.

We conclude that the given equation has:

- one solution if $\frac{pq-1}{q-p}$ is not an integer number;
- two solutions if $\frac{pq-1}{q-p}$ is an integer number but $\frac{q-1}{q-p}$ is not;
- four solutions if $\frac{q-1}{q-p}$ is an integer number.

Observation. A shorter path to the problem's conclusion is to notice that (1) is equivalent to

$$(pq - (q-p)x)(pq + (q-p)y) = p^2q^2.$$

Thus, we have

$$pq - (q-p)\,x = a$$
$$pq + (q-p)\,y = b,$$

where a and b are positive divisors of p^2q^2, such that $a < b$ and $ab = p^2q^2$. Since this implies $a < pq < b$, we have only four candidates for (a, b) : $\left(1, p^2q^2\right)$, $\left(p, pq^2\right)$, $\left(q, p^2q\right)$, and $\left(p^2, q^2\right)$. The solution ends as the previous one.

Appendix

The Euler Line

Let ABC be a triangle and let O, G, and H be its circumcenter, centroid, and orthocenter, respectively. If ABC is not equilateral, then the points O, G, and H are distinct and lie on the same line, called the Euler line (see Figure A.1).

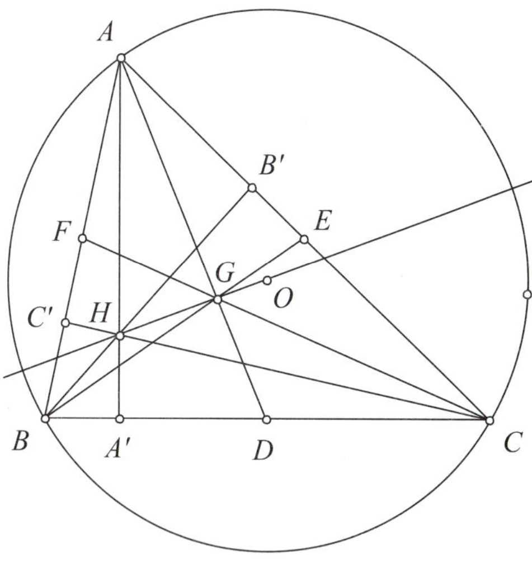

Figure A.1

Proof. Consider the following three lines: the median from A, the altitude from A and the perpendicular bisector of the side BC. It is not difficult to see that if two of these lines coincide, then all three coincide and the triangle ABC is isosceles. Therefore, if two of the points G, H, and O coincide, then all three coincide and the triangle ABC is equilateral.

Let A'' be a point on the circumcircle such that AA'' is a diameter. We prove that the quadrilateral $HBA''C$ is a parallelogram (see Figure A.2). Indeed, since AA'' is a diameter, the angles $\angle ACA''$ and $\angle ABA''$ are right angles. It follows that CA'' is perpendicular to AC, and hence it is parallel to HB. Similarly, BA'' is parallel to HC, therefore $HBA''C$ is a parallelogram.

This implies that the midpoint D of the side BC coincides with the midpoint of HA''. In the triangle AHA'', OD is a midline, so it is parallel to AH and $AH = 2OD$. Suppose the median AD and the line OH intersect at G'. Then the triangles ODG' and HAG' are similar and $G'A = 2G'D$. We deduce that G' is a point on the median AD such that

$$\frac{G'D}{G'A} = \frac{1}{2},$$

that is, G' coincides with G, the centroid of ABC. It follows that O, G, and H are collinear and that $GH = 2OG$.

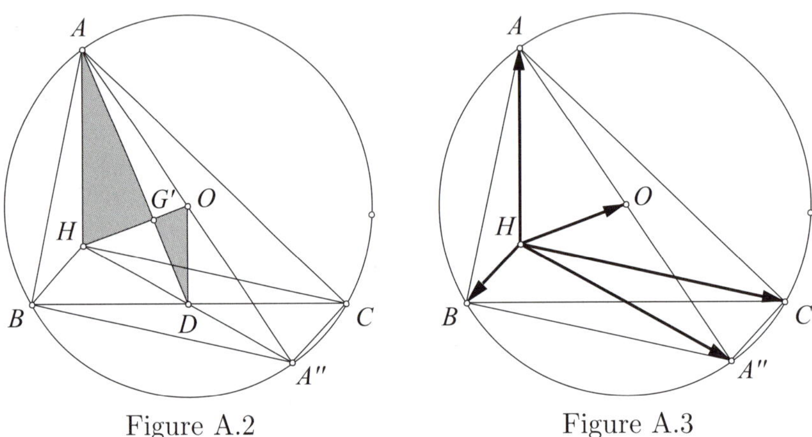

Figure A.2 Figure A.3

Another proof of the collinearity of the three points uses vectors. Since $HBA''C$ is a parallelogram, we have (see Figure A.3)

$$\overline{HB} + \overline{HC} = \overline{HA''},$$

and hence

$$\overline{HA} + \overline{HB} + \overline{HC} = \overline{HA} + \overline{HA''} = 2\overline{HO}.$$

This rewrites as

$$\overline{HO} + \overline{OA} + \overline{HO} + \overline{OB} + \overline{HO} + \overline{OC} = 2\overline{HO},$$

or

$$\overline{OA} + \overline{OB} + \overline{OC} = \overline{OH}. \tag{1}$$

On the other hand, it is known that for any point M we have

$$\overline{MA} + \overline{MB} + \overline{MC} = 3\overline{MG},$$

hence

$$\overline{OA} + \overline{OB} + \overline{OC} = 3\overline{OG}.$$

The equality

$$\overline{OH} = 3\overline{OG}$$

proves that the points O, G, and H are collinear.
Observation. Equality (1) shows that in the complex plane with origin at the circumcenter O, the affix of the orthocenter is

$$h = a + b + c,$$

where a, b, c are the affixes of the points A, B, and C.

The Nine Point Circle

In any triangle, the midpoints of the sides, the feet of the altitudes, and the midpoints of the segments joining the orthocenter to the vertices lie on the same circle (the nine point circle). The center of this circle is the midpoint of the segment joining the orthocenter and the circumcenter (see Figure A.4).

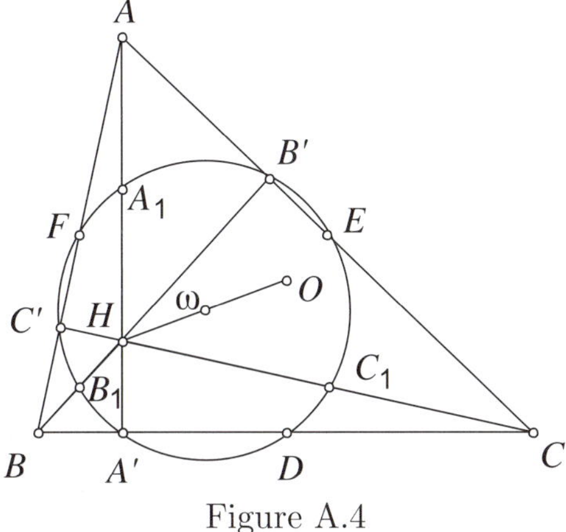

Figure A.4

Proof. Consider a triangle ABC and let D, E, and F be the midpoints of the sides BC, CA, and AB, respectively. Let A' be the foot of the altitude dropped from A (see Figure A.5). The quadrilateral $A'DEF$ is a trapezoid, since EF is parallel to BC. Moreover,

$$DE = \frac{AB}{2} = A'F$$

(in the right triangle ABA', the length of the median $A'F$ is half the length of the hypothenuse). Therefore, $A'DEF$ is an isosceles trapezoid, hence a cyclic one. We deduce that the point A' lies on the circumcircle of the triangle DEF. In the same way, we obtain that the two other feet of the altitudes, B' and C', lie on this circle as well.

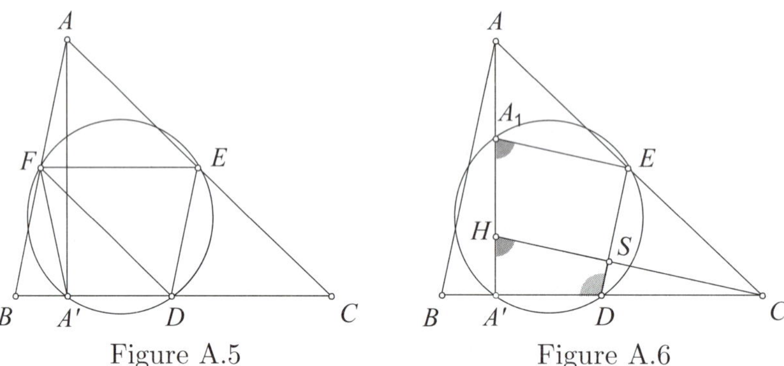

Figure A.5 Figure A.6

Let A_1 be the midpoint of the segment HA, where H is the orthocenter of the triangle ABC. In the triangle AHC, A_1E is a midline, so that A_1E is parallel to HC. It follows that $\angle HA_1E = \angle A'HC$. But the quadrilateral $HA'DS$ is cyclic (HS is perpendicular to DE and HA' is perpendicular to $A'D$), therefore

$$\angle A'HC = 180^\circ - \angle A'DS.$$

It follows that

$$\angle HA_1E = 180^\circ - \angle A'DS,$$

therefore the quadrilateral A_1EDA' is cyclic, too. This means that the point A_1 lies on the circumcircle of the triangle EDA', which is the same with the circumcircle of DEF. Similarly, we can prove that the two other midpoints of the segments HB and HC lie on the same circle.

Finally, we prove that the center of this circle is the midpoint ω of the segment OH, where O is the circumcenter of ABC.

We know (see The Euler line) that OD and HA are parallel and $HA = 2OD$. It follows that $OD = HA_1$, therefore $ODHA_1$ is a parallelogram (see Figure A.7). It follows that the midpoint ω of the segment OH coincides with the midpoint of the segment A_1D.

But the angle $\angle A_1A'D$ is a right angle, so that A_1D is a diameter of the nine point circle. We conclude that the point ω is indeed its center.

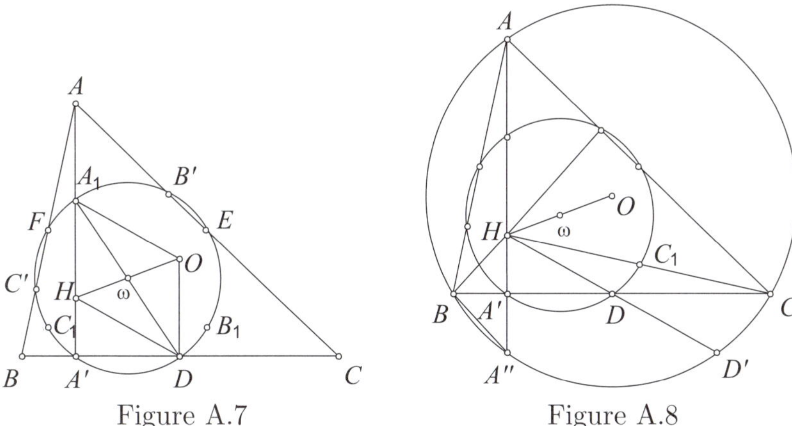

Figure A.7 Figure A.8

Another proof uses geometric transformations. Let A'' be the point of intersection between the line HA and the circumcircle of ABC (see Figure A.8).

We have

$$\angle A''BC = \angle A''AC = 90^\circ - \angle ACB = \angle CBB'.$$

It follows that BA' is an angle bisector in the triangle HBA''. But BA' is also an altitude in the same triangle. Therefore HBA'' is isosceles, hence A' is the midpoint of the segment HA''. If D' is a point on the circumcircle such that AD' is a diameter, we know (see The Euler line) that D is the midpoint of HD'. Because C_1 is the midpoint of HC, we observe that a homothety with center H and ratio $\frac{1}{2}$ sends the points A'', D', and C to the points A', D, and C_1.

It follows that A', D, and C_1 lie on the image through the homothety of the circumcircle of ABC, which is a circle with center ω, the midpoint of HO.

Euler Triangle Formula

Let ABC be an arbitrary triangle, O its circumcenter and I its incenter. If the circumradius and the inradius are denoted by R and r, then

$$OI^2 = R^2 - 2Rr.$$

Proof. Suppose the angle bisector of $\angle A$ intersects the circumcircle at A' and let E be a point on the circumcircle such that $A'E$ is a diameter (see Figure A9). If D is the projection of the incenter on the side AC then $ID = r$ and the triangles AID and $EA'C$ are similar, hence

$$\frac{AI}{ID} = \frac{EA'}{A'C}. \tag{1}$$

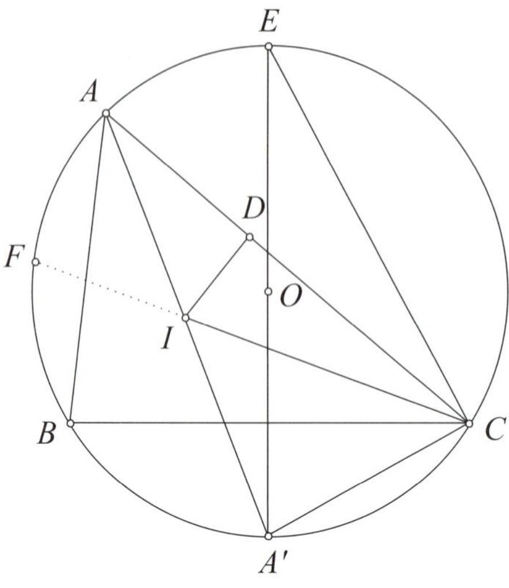

Figure A.9

Observe that the triangle $IA'C$ is isosceles. Indeed, $\angle A'IC$ is an interior angle for the circumcircle, so

$$\begin{aligned}\angle A'IC &= \frac{1}{2}\left(\text{arc}AF + \text{arc}CA'\right)\\ &= \frac{1}{2}\left(\text{arc}FB + \text{arc}BA'\right)\\ &= \angle A'CI.\end{aligned}$$

It follows that $A'C = IA'$, therefore (1) becomes

$$\frac{AI}{ID} = \frac{EA'}{IA'},$$

or

$$AI \cdot IA' = 2R \cdot r.$$

The power of a point theorem applied to I gives

$$AI \cdot IA' = R^2 - OI^2,$$

hence

$$OI^2 = R^2 - 2R \cdot r.$$

Another possible approach is by computations. It is known that for any point M in the plane of the triangle, the following equality holds:

$$\overline{MI} = \frac{a\overline{MA} + b\overline{MB} + c\overline{MC}}{a+b+c}.$$

Applying this for $M = O$ and squaring yields

$$OI^2 = \frac{1}{(a+b+c)^2}\left(a\overline{OA} + b\overline{OB} + c\overline{OC}\right)^2$$
$$= \frac{1}{(a+b+c)^2}\left(R^2\sum a^2 + 2\sum ab\cdot\overline{OA}\cdot\overline{OB}\right).$$

But

$$\overline{OA}\cdot\overline{OB} = R^2\cos\angle 2C$$
$$= R^2\left(1 - 2\sin^2\angle C\right)$$
$$= R^2\left(1 - \frac{c^2}{2R^2}\right)$$
$$= R^2 - \frac{1}{2}c^2.$$

We obtain

$$OI^2 = \frac{1}{(a+b+c)^2}\left(R^2\left(\sum a^2 + 2\sum ab\right) - abc\,(a+b+c)\right)$$
$$= \frac{1}{(a+b+c)^2}\left(R^2\,(a+b+c)^2 - abc\,(a+b+c)\right)$$
$$= R^2 - \frac{abc}{a+b+c}.$$

Finally, replacing $abc = 4RS$ and $a+b+c = \frac{2S}{r}$ gives the desired result.
Observation. Euler's formula obviously implies the well known inequality

$$R \geq 2r$$

(also known as Euler's inequality) which holds in every triangle.

Leibniz's Relation

Let ABC be a triangle and let O and G be its circumcenter and its centroid. If R is the radius of the circumcircle, then

$$OG^2 = R^2 - \frac{1}{9}\left(a^2 + b^2 + c^2\right). \qquad (1)$$

Proof. It is known that

$$3\overline{OG} = \overline{OA} + \overline{OB} + \overline{OC}.$$

Squaring this equality yields

$$9OG^2 = \sum OA^2 + 2\sum \overline{OA} \cdot \overline{OB} = 3R^2 + 2\sum \overline{OA} \cdot \overline{OB}.$$

We have

$$\overline{OA} \cdot \overline{OB} = OA \cdot OB \cdot \cos \angle AOB = R^2 \cos 2C$$

and using a trigonometric formula and the sine law we obtain

$$R^2 \cos 2C = R^2 \left(1 - 2\sin^2 C\right) = R^2 \left(1 - \frac{a^2}{2R^2}\right) = R^2 - \frac{1}{2}a^2.$$

It follows that

$$9OG^2 = 3R^2 + 2\left(3R^2 - \frac{1}{2}\sum a^2\right) = 9R^2 - \sum a^2.$$

Dividing by 9 gives the desired result.
Observation. As proved in "The Euler Line", we have

$$OH = 3OG,$$

therefore (1) becomes

$$\frac{1}{9}OH^2 = R^2 - \frac{1}{9}\left(a^2 + b^2 + c^2\right),$$

or

$$OH^2 = 9R^2 - \left(a^2 + b^2 + c^2\right).$$

This can also be proved by squaring the equality

$$\overline{OH} = \overline{OA} + \overline{OB} + \overline{OC}.$$

The Incircle and the Excircles of a Triangle

Given the triangle ABC, there exists a circle tangent to its sides (called the *incircle*), centered at a point I lying in the interior of the triangle. The point I (the *incenter*) is equally distanced from the triangle's sides and hence it is the intersection point of the interior angle bisectors (see Figure A.10).

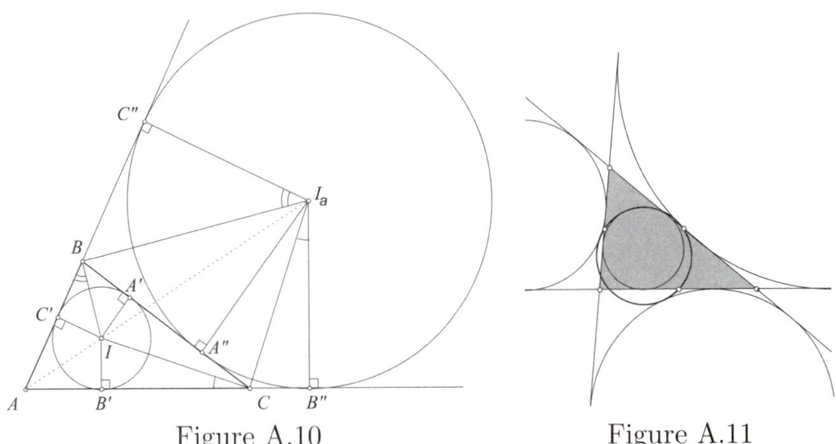

Figure A.10

Figure A.11

Three other points are also equally distanced from the triangle's sides. The internal angle bisector of $\angle A$ and the exterior angle bisectors of $\angle B$ and $\angle C$ intersect at I_a, the center of one of the *excircles* of triangle ABC. The points I_b and I_c are similarly defined.

One of the most beautiful problem in elementary geometry is Feuerbach's theorem which states that the nine point circle is tangent to the incircle and to all three excircles of a triangle (see Figure A.11).

Since the tangents to a circle from a point are equal in length, we can denote $x = AB' = AC'$, $y = BC' = BA'$, $z = CA' = CB'$. Then

$$\begin{aligned} AB &= c = x + y \\ AC &= b = x + z \\ BC &= a = y + z. \end{aligned}$$

Solving for x, y, z, we obtain

$$x = s - a,\ \ y = s - b,\ \ z = s - c,$$

where $s = \frac{1}{2}(a + b + c)$ is the semiperimeter.

Observation. We can see that the positive numbers a, b, c are the side lengths of a triangle if and only if there exist positive numbers x, y, z such that $a = y + z$, $b = x + z$, $c = x + y$. This can be very useful in algebraic manipulations (e.g. inequalities) involving the lengths of a triangle's sides.

If r denotes the inradius of the triangle, then, in the triangle AIB', we have

$$r = (s - a)\tan\frac{A}{2}.$$

Similarly,

$$r = (s-b)\tan\frac{B}{2} = (s-c)\tan\frac{C}{2}.$$

Now, denote $AB'' = AC'' = x'$, $BC'' = BA'' = y'$, and $CA'' = CB'' = z'$ (see Figure A.10). Then

$$\begin{aligned} 2s &= AB + BC + CA \\ &= AC'' - BC'' + BA'' + CA'' + AB'' - CB'' \\ &= AC'' + AB'' \\ &= 2x', \end{aligned}$$

and hence $x' = s$. It follows that $y' = AC'' - AB = s - c$, and $z' = s - b$.

The Eisenstein Criterion

Let $f(X) = a_0 + a_1X + \ldots + a_nX^n$ be a polynomial with integer coefficients. If there exists a prime number p such that the coefficients $a_0, a_1, \ldots, a_{n-2}$ and a_{n-1} are divisible by p, a_n is not divisible by p and a_0 is not divisible by p^2, then f cannot be written as a product of two nonconstant integer polynomials.
Proof. Suppose, by way of contradiction, that $f = gh$, where g and h are nonconstant polynomials with integer coefficients. If

$$\begin{aligned} g(X) &= b_0 + b_1X + \ldots + b_sX^s, \\ h(X) &= c_0 + c_1X + \ldots + c_tX^t, \end{aligned}$$

with $s \le t$ then the equality $f = hg$ gives

$$\begin{aligned} a_0 &= b_0c_0 \\ a_1 &= b_0c_1 + b_1c_0 \\ a_2 &= b_0c_2 + b_1c_1 + b_2c_0 \\ &\vdots \\ a_s &= b_0c_s + b_1c_{s-1} + \ldots + b_{s-1}c_1 + b_sc_0 \\ &\vdots \\ a_t &= b_0c_t + b_1c_{t-1} + \ldots + b_sc_{t-s} \\ &\vdots \\ a_n &= b_sc_t. \end{aligned}$$

Because a_0 is divisible by p but not by p^2, it follows that exactly one of the numbers b_0 and c_0 is divisible by p. Suppose that p divides b_0 and does not divide c_0.

Since p divides $a_1 = b_0c_1 + b_1c_0$ and divides b_0, it follows that p divides b_1c_0; but p does not divide c_0, therefore p divides b_1.

Now, p divides $a_2 = b_0c_2 + b_1c_1 + b_2c_0$ and divides b_0 and b_1, therefore p divides b_2c_0; since p does not divide c_0, it results that p divides b_2.

In the same way, we obtain successively that p divides $b_3, \ldots, b_{s-1}$, and b_s. But then p divides $b_sc_t = a_n$, which is a contradiction.

As an application of the Eisenstein criterion, let us prove the result used in the solution of the problem 18.2.

Lemma. If p is a prime number, then the polynomial

$$f(X) = X^{p-1} + X^{p-2} + \ldots + X + 1$$

cannot be written as a product of two nonconstant integer polynomials.

Proof. We have

$$f(X) = \frac{X^p - 1}{X - 1},$$

hence the polynomial $g(X) = f(X+1)$ can be written as

$$g(X) = \frac{(X+1)^p - 1}{X} = X^{p-1} + \binom{p}{1}X^{p-2} + \ldots + \binom{p}{p-2}X + \binom{p}{p-1}.$$

Observe that we can apply the Eisenstein criterion for the prime number p. Indeed, for each k, $1 \le k \le p-1$, p divides $\binom{p}{k}$, since

$$k\binom{p}{k} = p\binom{p-1}{k-1},$$

and p^2 does not divide $\binom{p}{p-1} = p$.

The Rearrangements Inequality

Let $n \ge 2$ be a positive integer and let $a_1 < a_2 < \ldots < a_n$, $b_1 < b_2 < \ldots < b_n$ be two ordered sequences of real numbers. Then, considering all the sums of the form

$$S(\sigma) = a_1b_{\sigma(1)} + a_2b_{\sigma(2)} + \ldots + a_nb_{\sigma(n)},$$

where σ is a permutation of the numbers $1, 2, \ldots, n$, the maximal one is

$$a_1b_1 + a_2b_2 + \ldots + a_nb_n,$$

and the minimal one is

$$a_1b_n + a_2b_{n-1} + \ldots + a_nb_1.$$

Indeed, let σ be a permutation for which $S(\sigma)$ is maximal. (Such a permutation exists since the number of possible sums is finite.) Suppose, by way of contradiction, that one can find i, j, with $1 \leq i < j \leq n$, such that $\sigma(i) > \sigma(j)$. Now, the new permutation σ', defined by

$$\sigma'(k) = \begin{cases} \sigma(k), & \text{for } k \neq i, j, \\ \sigma(j), & \text{for } k = i, \\ \sigma(i), & \text{for } k = j. \end{cases}$$

We have

$$\begin{aligned} S(\sigma') - S(\sigma) &= a_i b_{\sigma(j)} + a_j b_{\sigma(i)} - a_i b_{\sigma(i)} - a_j b_{\sigma(j)} \\ &= (a_i - a_j)\left(b_{\sigma(j)} - b_{\sigma(i)}\right) > 0, \end{aligned}$$

since $a_i < a_j$ and $b_{\sigma(j)} < b_{\sigma(i)}$. This implies $S(\sigma') > S(\sigma)$, contradicting thus the maximality of $S(\sigma)$.

In a similar manner one can prove that $a_1 b_n + a_2 b_{n-1} + \ldots + a_n b_1$ is the minimal sum.

Observations. 1. If we replace the initial conditions with the less restrictive ones

$$a_1 \leq a_2 \leq \ldots \leq a_n, \; b_1 \leq b_2 \leq \ldots \leq b_n,$$

the conclusion still holds, only in this case there might exist more than one maximal (minimal) sum. Just consider the extreme case $a_1 = a_2 = \ldots = a_n$.

2. If the given sequences have opposite monotonies, then the first sum is minimal and the second one maximal.

Young's Inequality

Let $f : [0, +\infty) \to \mathbb{R}$ be a continuous, strictly increasing function, with $f(0) = 0$ and $\lim\limits_{x \to +\infty} f(x) = +\infty$. Then

$$\int_0^a f(x)\,dx + \int_0^b f^{-1}(x)\,dx \geq ab,$$

for any positive numbers a and b.

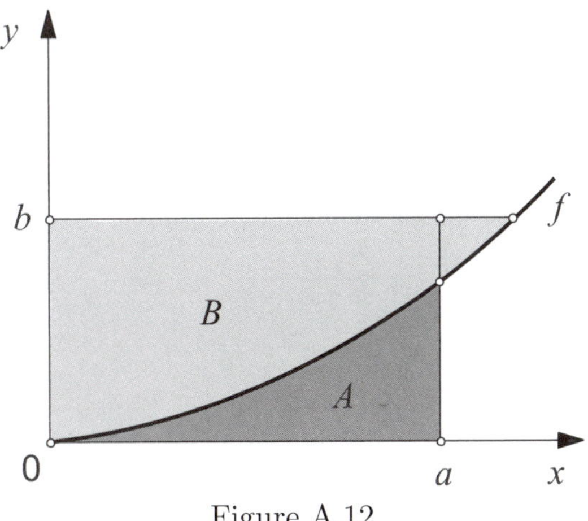

Figure A.12

The inequality has a geometric interpretation. The integral $\int_0^a f(x)\,dx$ equals the area of the surface bounded by the graph of f, the Ox axis and the line $x = a$ (denoted by A in Figure A.12). The integral $\int_0^b f^{-1}(x)\,dx$ equals the area of the surface bounded by the graph of f, the Oy axis and the line $y = b$ (denoted by B in Figure A.12).

It is easy to see that the sum of the two areas is no less than the area of the rectangle with side lengths a and b. Moreover, the equality holds if and only if $f(a) = b$.

Glossary

AM-GM inequality
For any positive numbers $x_1, x_2, \ldots, x_n$ we have

$$\frac{x_1 + x_2 + \ldots + x_n}{n} \geq \sqrt[n]{x_1 x_2 \ldots x_n},$$

with equality if and only if $x_1 = x_2 = \ldots = x_n$.

Angle bisector theorem
If D is a point on the side BC of the triangle ABC such that AD bisects the angle $\angle A$ then

$$\frac{BD}{DC} = \frac{AB}{AC}.$$

Bijective Function
A function which is both one to one and onto.

Cauchy Schwartz inequality
If $a_1, a_2, \ldots, a_n, b_1, b_2, \ldots, b_n$ are nonzero real numbers, then

$$\left(a_1^2 + a_2^2 + \ldots + a_n^2\right)\left(b_1^2 + b_2^2 + \ldots + b_n^2\right) \geq \left(a_1 b_1 + a_2 b_2 + \ldots + a_n b_n\right)^2,$$

with equality if and only if $\frac{a_1}{b_1} = \frac{a_2}{b_2} = \ldots = \frac{a_n}{b_n}$.

Centroid of a triangle
Point of intersection of the medians.

Ceva's theorem
Let AA', BB', and CC' be three cevians of triangle ABC. Then AA', BB', and CC' are concurrent if and only if

$$\frac{A'B}{A'C} \cdot \frac{B'C}{B'A} \cdot \frac{C'A}{C'B} = 1.$$

Ceva's theorem (trigonometric form)
Let AA', BB', and CC' be three cevians of triangle ABC. Then AA', BB', and CC' are concurrent if and only if

$$\frac{\sin\angle A'AB}{\sin\angle A'AC} \cdot \frac{\sin\angle B'BC}{\sin\angle B'BA} \cdot \frac{\sin\angle C'CA}{\sin\angle C'CB} = 1.$$

Cevian
Any segment joining the vertex of a triangle to a point on the opposite side.

Circumcenter
Center of the circumscribed circle.

Circumcircle
Circumscribed circle.

Convex function
The function $f : I \rightarrow \mathbb{R}$, where I is an interval, is convex (concave up) if for any $x_1, x_2 \in I$ and any $\lambda \in [0,1]$ we have

$$\lambda f(x_1) + (1-\lambda) f(x_2) \geq f(\lambda x_1 + (1-\lambda) x_2).$$

Convex quadrilateral
The quadrilateral $ABCD$ is convex if the line segments AC and BD intersect (or, equivalently, if all its interior angles are less than 180°).

Cosine law
In the triangle ABC we have

$$BC^2 = AB^2 + AC^2 - 2AB \cdot AC \cos \angle BAC.$$

Cyclic polygon
Polygon that can be inscribed in a circle.

Eisenstein's criterion
Let $f = a_0 + a_1X + \ldots + a_nX^n$ be an integer polynomial. If p is a prime number such that p divides $a_0, a_1, \ldots a_{n-1}$, p does not divide a_n and p^2 does not divide a_0, then f cannot be written as a product of two (nonconstant) integer polynomials.

Euler Line
The line which passes through the circumcenter, orthocenter and centroid of a triangle.

Fermat's little theorem
If p is a prime number, then $a^p \equiv a \pmod{p}$ for any integer a. Furthermore, if p does not divide a, then $a^{p-1} \equiv 1 \pmod{p}$

HM-AM inequality
For any positive numbers $x_1, x_2, \ldots, x_n$ we have

$$\frac{n}{\frac{1}{x_1} + \frac{1}{x_2} + \ldots + \frac{1}{x_n}} \leq \frac{x_1 + x_2 + \ldots + x_n}{n},$$

with equality if and only if $x_1 = x_2 = \ldots = x_n$.

Heron's formula
If a, b, c are the side lengths of a triangle, S denotes its area, and $p = \frac{a+b+c}{2}$ is the semiperimeter, then

$$S = \sqrt{p(p-a)(p-b)(p-c)}.$$

Homothety
A homothety of center O and ratio r is a transformation that maps each point P in the plane to a point P' such that $\overline{OP'} = r\overline{OP}$.

Incenter
Center of the inscribed circle.

Incircle
Inscribed circle.

Inradius
The radius of the incircle.

Inversion
If O is a point in the plane and ρ is a positive real number, the inversion through O with radius ρ is a transformation which maps every point $P \neq O$ to a point $P' \in (OP$ such that $OP \cdot OP' = \rho^2$.

Isogonal conjugate
The isogonal conjugate X^* of a point X in the plane of the triangle ABC is constructed by reflecting the lines AX, BX, and CX about the angle bisectors at A, B, and C. The three reflected lines then concur at the isogonal conjugate X^*.

Jensen's inequality
If the function $f : I \to \mathbb{R}$, where I is an interval, is convex (concave up) then for any $x_1, x_2, \ldots, x_n \in I$ and any $\lambda_1, \lambda_2, \ldots, \lambda_n \in [0,1]$ such that $\sum \lambda_i = 1$, we have

$$\lambda_1 f(x_1) + \lambda_2 f(x_2) + \ldots + \lambda_n f(x_n) \geq f(\lambda_1 x_1 + \lambda_2 x_2 + \ldots + \lambda_n x_n).$$

The inequality is reversed for a concave down function.

Median theorem
If AD is a median in the triangle ABC then

$$AD^2 = \frac{2(AB^2 + AC^2) - BC^2}{4}.$$

Menelaus' theorem
The points M, N, and P are chosen on the lines AB, BC, and CA such that

either exactly one or all three lie outside the line segments AB, BC, and CA, respectively. Then M, N, P are collinear if and only if

$$\frac{MA}{MB} \cdot \frac{NB}{NC} \cdot \frac{PC}{PA} = 1.$$

De Moivre's formula
For any angle a and any integer n,

$$(\cos a + i \sin a)^n = \cos na + i \sin na.$$

The nine point circle
The circle which passes through the midpoints of the sides, the feet of the altitudes, and the midpoints of the segments joining the orthocenter to the vertices of a given triangle.

One-to-one function
A function $f : A \to B$ such that if $x \neq y$ then $f(x) \neq f(y)$.

Onto function
A function $f : A \to B$ such that for any $y \in B$ there exists at least one $x \in A$ such that $f(x) = y$.

Orthocenter
The point of intersection of the altitudes of a triangle.

Power of a point
If P lies inside a circle $\mathcal{C}$ and the chords AB, $A'B'$ pass through P, then

$$PA \cdot PB = PA' \cdot PB' = R^2 - OP^2,$$

where O and R are the circumcenter and the radius of P. If P lies outside $\mathcal{C}$ then

$$PA \cdot PB = PA' \cdot PB' = OP^2 - R^2.$$

The number $\rho(P) = OP^2 - R^2$ is called the power of the point P with respect to the circle $\mathcal{C}$.

Root of unity
Solution to the equation $z^n - 1 = 0$.

Sine law
If a, b, c are the side lengths of the triangle ABC and R is its circumradius, then

$$\frac{a}{\sin \angle A} = \frac{b}{\sin \angle B} = \frac{c}{\sin \angle C} = 2R.$$

Symmetry center
The point O is a symmetry center of a figure F if for any point $M \in F$, there exists $M' \in F$ such that O is the midpoint of the line segment MM'.

Viète's relations

If $x_1, x_2, \ldots, x_n$ are the roots of the polynomial

$$P(X) = a_0 + a_1 X + \ldots + a_n X^n,$$

then the following equalities hold:

$$\begin{aligned}
x_1 + x_2 + \ldots + x_n &= -\frac{a_{n-1}}{a_n} \\
x_1 x_2 + x_1 x_3 + \ldots + x_{n-1} x_n &= \frac{a_{n-2}}{a_n} \\
&\vdots \\
x_2 x_3 \ldots x_n + x_1 x_3 \ldots x_n + \ldots + x_1 x_2 \ldots x_{n-1} &= (-1)^{n-1} \frac{a_1}{a_n} \\
x_1 x_2 \ldots x_n &= (-1)^n \frac{a_0}{a_n}.
\end{aligned}$$

Index of Notations

$\mathbb{Z}$	the set of integers
$\mathbb{Q}$	the set of rational numbers
$\mathbb{R}$	the set of real numbers
$\mathbb{C}$	the set of complex numbers
$\mathbb{Z}_n$	the ring of residues modulo n
$\mathbb{F}_p$ (or $\mathbb{Z}_p$)	the field of residues modulo p
$\|A\|$	the number of elements of the finite set A
$[a, b]$	the set of real numbers x such that $a \leq x \leq b$
(a, b)	the set of real numbers x such that $a < x < b$
AB	the line or the segment AB; also the length of the segment AB
$(AB$	the ray (halfline) AB
$\overline{AB}$	the vector AB
$[F]$	the area of the figure F
$\lfloor x \rfloor$	the integer part of the real number x

References

1. Ioan Tomescu et al., *Olimpiadele Balcanice de Matematică*, Editura Gil, Zalău, 1996 (Romanian).

2. V. Iankovici, Z. Kadelburg, P. Mladenovici, *Mejunarodne i Balkanske Matematicke Olimpiade*, Beograd, 1996 (Serbian).

3. M. Becheanu, B. Enescu, *Romanian Mathematical Competitions*, Bucharest, 1999.

4. M. Becheanu, M. Baluna, B. Enescu, *Romanian Mathematical Competitions*, Bucharest, 2000.

5. Gazeta Matematica, (journal) 1984-2012.

6. Titu Andreescu, Bogdan Enescu, *Mathematical Olympiad Treasures, Second edition,* Birkhäuser, 2011.

7. *www.mathlinks.ro*

Other Books from XYZ Press

1. Andreescu, T.; Kisacanin, B., *Math Leads for Mathletes, a rich resource for young math enthusiasts, parents, teachers, and mentors*, Book 1, 2014.
2. Andreescu, T., *Mathematical Reflections - two more years*, 2014.
3. Andreescu, T.; Ganesh, A., *108 Algebra Problems from the AwesomeMath Year-Round Program*, 2014.
4. Andreescu, T.; Rolinek, M.; Tkadlec, J., *107 Geometry Problems from the AwesomeMath Year-Round Program*, 2013.
5. Andreescu, T.; Rolinek, M.; Tkadlec, J., *106 Geometry Problems from the AwesomeMath Summer Program*, 2013.
6. Andreescu, T., *105 Algebra Problems from the AwesomeMath Summer Program*, 2013.
7. Andreescu, T.; Kane, J., *Purple Comet Math Meet! - the first ten years*, 2013.
8. Andreescu, T.; Dospinescu, G., *Straight from the Book*, 2012.
9. Andreescu, T.; Boreico, I.; Mushkarov, O.; Nikolov, N., *Topics in Functional Equations*, 2012.
10. Andreescu, T., *Mathematical Reflections - the next two years*, 2012.
11. Andreescu, T., *Mathematical Reflections - the first two years*, 2011.
12. Andreescu, T.; Dospinescu, G., *Problems from the Book*, Second edition, 2010.

Printed by "Combinatul Poligrafic"
Com. nr. 80247